McGraw-Hill Book Company

Publishers of Books for

Electrical World	The Engineering and Mining Journal
Engineering Record	Engineering News
Railway Age Gazette	American Machinist
Signal Engineer	American Engineer
Electric Railway Journal	Coal Age
Metallurgical and Chemical Engineering	Power

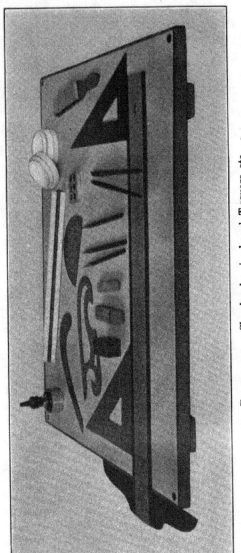

FRONTISPIECE.—Showing drawing board, T-square, etc.

ELEMENTARY MECHANICAL DRAWING

THEORY AND PRACTICE

WITH CHAPTERS ON

GEOMETRICAL DRAWING, MENSURATION

AND

REPRODUCTION OF DRAWINGS

A TEXT-BOOK FOR TECHNICAL, SECONDARY, TRADE AND VOCATIONAL SCHOOLS

BY

CHARLES WILLIAM WEICK, B. Sc.

ASSISTANT PROFESSOR OF DRAWING AND DESIGN
TEACHERS COLLEGE, COLUMBIA UNIVERSITY
IN THE CITY OF NEW YORK

FIRST EDITION

McGRAW-HILL BOOK COMPANY, Inc.
239 WEST 39TH STREET, NEW YORK
6 BOUVERIE STREET ,LONDON, E. C.
1915

THE MAPLE PRESS YORK PA

PREFACE

This work aims to provide an elementary text-book for class use on the modern conventions, theory, and practice of mechanical drawing. In scope and purpose it is designed to offer a basic treatment of the subject. Beginning with the elements it develops instruction in regular progress till it covers the fundamental training needed for general practice in the drafting office.

Its object is to cover both theory and practice. The treatment of theory in the text is as brief as clearness permits, but it is expressed with all necessary fulness and elaboration by the aid of many illustrations in which the theory is implicit. The conventions of the drawing-room are duly explained. Emphasis, however, is laid on practice, for mechanical drawing is essentially a practical art, and copious examples and problems exemplify and enforce the theory at every stage of the instruction.

The fundamentals necessary for the intelligent study and prosecution of a course in mechanical drawing are contained in the Chapters I–VI, preceding the examples and problems. These chapters when used are best taken up in connection with the course in practice as it develops. The chapter on geometrical drawing contains many principles useful to the draftsman in practice. The student should review this chapter from time to time till he is familiar with these principles and can avail himself of their aid in the execution of appropriate problems in the course in drawing. The chapter on mensuration affords helpful material of instruction for such classes in mechanical drawing as include mensuration in the work of their course.

Instruction in practice given through the examples and problems, with the exception of those intended for preliminary instrument practice and those illustrating simple projection, are concrete and practical exercises. The problems are arranged in sequence to form a graded series beginning with the simplest object consisting of one piece and ending with an object consisting of a number of pieces. The series of problems constitutes two alternative parallel courses, one intended for those interested in wood construction, the other for those interested in metal work.

v

The allotted time stated for the solution of each problem is based on the average student's ability, and furnishes a standard of speed in the work of the class.   Each example is accompanied by three problems of increasing difficulty; the teacher can make a selection to suit students of different ability and so keep the class together in their general advance through the subject.

In practice all problems of mechanical drawing involve the representation of an object and its parts on a sheet of paper. The placing and grouping of the various views of a drawing are important for the sake of clearness, convenience in use, or general neatness and workmanlike finish.   The layout drawing, shown with each example, directs constant attention to this feature of good draftmanship, and offers a method by which the student may steadily gain facility in good composition.

The chapter on the reproduction of drawings is given to familiarize the student with the method of reproducing drawings in quantity, from one original, for commercial purposes.

Drawing from models has a certain value in the study of mechanical drawing, but never without the association of theory. The danger which besets drawing from models is that the student may acquire the habit of mere mechanical imitation.   Such a habit is fatal to competency.   It is only when the draftsman understands his subject analytically that his work becomes intelligent and creative.   Such use as can be made of models may be safely left to the discretion of the teacher.

Traditional problems involving lines of intersection and developments of surfaces, and also problems in isometric drawing, belong to specialized branches of mechanical drawing and do not fall within the scope of the present work.   The instruction offered by this volume aims at the production of what are commonly known as working drawings.   The theory of the text can, however, be used in connection with any course which contains such problems.

In preparing the manuscript and drawings for this book the author has received many valuable suggestions from Mr. Frank C. Panuska, instructor of mechanical drawing, Teachers College; Mr. Arthur Hopper, instructor of mechanical drawing, Rochelle Park High School, Elizabeth, N. J., and Mr. Charles H. Meeker, whose help is gratefully acknowledged.

NEW YORK,
April,1915

CONTENTS

CHAPTER VII

CHAPTER VIII

CHAPTER IX

CHAPTER X

INTRODUCTION

Mechanical drawing is the art of exact drawing by means of drawing instruments of precision. It differs from free-hand drawing because it uses instruments and in general aims to reproduce the object not as it appears to the eye in perspective but in its absolute relations in space. Mechanical Drawing is a generic title; specialized branches of mechanical drawing are known as architectural drawing, topographical drawing, machine design, etc. The elements which underlie or are involved in the art of mechanical drawing are: First, The Theory of Projection; Second, Geometrical Drawing, and Third, Mathematics.

The principles of projection are involved, since we aim to transfer objects having three dimensions to one plane of two dimensions (the drawing paper). The difficulty which usually attends the study of projection will be greatly lessened if a good method of instruction is followed. It is better, for instance, to begin with simple objects, than with complex ones. It is better to study minutely each separate detail, rather than several parts at once; studies of unit details can then be built up to cover the whole of a complex object. The student, moreover, should not be content with one study of the object but should make several studies of the same object in different positions. He should early acquire the habit of analyzing, from the draftsman's point of view, every object into its parts and details. This power of analysis once acquired saves time, ensures correctness, and is the secret of power in the handling of difficult and unusual problems.

Geometry is indispensable, and geometrical problems are included for reference in the present work. Every draftsman at some time or other has problems for the solution of which the T-square or triangle are quite inadequate; he must have recourse to geometric construction, or fail. It is true that often other means are more convenient than geometric methods, but it is likewise true that in many cases geometric methods are essential, and that by their use alone can absolute accuracy be attained. It is perhaps not going too far to say that geometric drawing holds the same place to the trained draftsman as the alphabet

does to the student in language. Geometric problems are given
in the present work with greater fulness than any single student
will probably require; he will choose and select according to his
needs; and the problems are varied to meet the needs of many
classes of students.

The accuracy that the practice of geometric drawing produces
is one of the prime qualities of the good draftsman. This accuracy is required in all modern industries which involve the use of
design, patterns and construction. Economy in production
depends on it. The draftsman's mistakes reproduced in costly
material by costly labor entail great waste. Even apparently
trivial errors may result in serious loss of time and material.

Since the purpose of mechanical drawing is to represent exactly the proportions of an object, the draftsman has frequent
recourse to mathematics for the accurate calculations of dimensions. In the practical work of the drawing-room the draftsman
must be able to make calculations of many kinds—areas, volumes,
and weights of machine parts, stresses and strains in parts of
buildings and machines.

The student should begin his study of mechanical drawing
with practical work on specific drawing problems. His study of
theory should not precede all drawing, but should fit itself into
the drawing course in parallel and cooperating development.

In beginning the study of every problem, the student should
study and master the layout of the example, and follow it as a
guide in placing the several views constituting the problem.
The analysis of the problem and the successful treatment of it,
depend to a great extent on the thoroughness with which the
student studies the layout drawing.

ELEMENTARY
MECHANICAL DRAWING

CHAPTER I

INSTRUMENTS AND MATERIALS

1. Introductory.

The instruments used in mechanical drawing may be divided into two distinct groups: first, the set of instruments, generally in a case, consisting of a compass, dividers, ruling pens, etc., and second, the drawing board, T-square, triangles, etc. To secure good results in mechanical drawing good instruments are necessary. Economy in the purchase of instruments is unwise, because drawings made with inferior instruments will be of inferior quality in technique. A good set of instruments with proper care will serve the student's needs almost indefinitely. The instruments of the second group, such as the drawing-board, etc., may, however, be of medium grade, if the cost of the very best is prohibitive.

The materials used may also be divided into two parts: first, such materials as drawing-paper, tracing-paper, tracing-cloth, and printing-paper; and second, pencils, erasers, inks, etc. The drawing-paper may be of medium grade; it should be either white or light tan in color. The tracing-paper or tracing-cloth and printing-paper, also the materials of the second part, should be of the very best quality; otherwise good results are impossible. To insure quality in the instruments and materials purchased, they should be procured only from reliable dealers in drawing supplies.

The instruments described in this chapter are divided into two sections, the first consisting of the minimum equipment which is necessary for a beginner to do commendable work. The appended prices are based on the price lists of dealers in New York City, and will doubtless stand as a fair average for

1

instruments of desirable grade everywhere. The exact price
for the same grade of instruments may, of course, vary slightly
in other localities.

The second group consists of what are known as special instruments. These do not generally form a part of a student's or
draftsman's outfit, but are supplied by the school or drawing
office. Advanced students and draftsmen prefer to purchase
many of them, as it is obviously impossible to do work of merit
with tools which are used promiscuously and often carelessly.

Before purchasing an outfit, it would be well for the beginner
to consult someone who is capable of judging the quality of
instruments and materials and enlist his help in making the
selection.

LIST OF INSTRUMENTS

2. Minimum Equipment.

A set of instruments in pocket case includes the following (see
Figs. 1 and 2):

1. Compass, 5–1/2″, with fixed needle-point leg, pencil leg, pen
   leg, and extension bar.
   Dividers, 5″, with hair-spring adjustment.
   Bow Pencil, 3″.
   Bow Pen, 3″.
   Bow Dividers, 3″.
   Ruling Pen, 6″.
   Ruling Pen, 5″.
   A good set complete, costs................................ $11.50
2. Drawing board, 20″ × 26″................................ 1.00
3. T-square, 25″ long, with celluloid edges..................... 1.00
4. 45° celluloid triangle, 7″................................. .30
5. 30° and 60° celluloid triangle, 9″......................... .30
6. Two celluloid curves...................................... .50
7. 12″ architects triangular scale, boxwood.................... .40
8. 3H pencil for lettering and figures......................... .08
9. 4H pencil for drawing..................................... .08
10. 5H pencil for drawing.................................... .08
11. Sand paper pencil pointer................................ .10
12. Ink and pencil eraser.................................... .05
13. Sponge rubber, or art gum............................... .10
14. Two penholders and pens for lettering..................... .12
15. Thumb tacks, one dozen................................. .10
16. Ink, one bottle, black, waterproof........................ .25
17. Soapstone.. .05
18. Chamois, small piece.................................... .10

Total... $16.11

3. Compass. (Fig. 3.)

The compass should be of light weight, about 6 inches long and fitted with pencil leg, pen leg and extension bar. The fixed leg should be provided with a removable needle point, one end of

Fig. 1.—Set of instruments in pocket case with flaps.

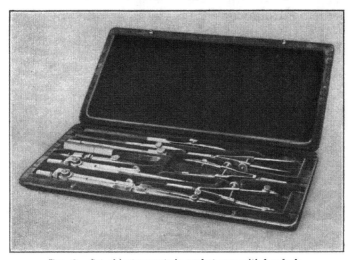

Fig. 2.—Set of instruments in pocket case with bar-lock.

which may have a long taper point. The other end must be provided with a very fine point, somewhat less than 1/16 inch long, with a shoulder which will prevent the needle from entering the paper too far, thereby enlarging the hole and tending to produce inaccurate results.

Use.—The compass is used for drawing circles, or parts of circles, with either pencil or ink. The pencil should be 4H or 5H, and sharpened to a chisel point on a piece of sand paper or a fine file. For ink work the pen leg is to be inserted into the compass and ink introduced as in the straight-line pen. The extension bar is to be inserted for drawing large circles.

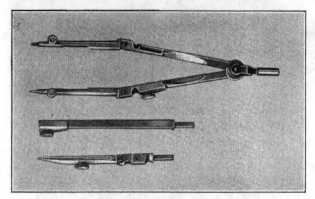

FIG. 3.—Compass—complete outfit.

4. Dividers. (Fig. 4.)

The dividers, like the compass, should be of light weight and about 5 inches long. It is desirable to have one of the legs provided with a hair-spring adjustment, for fine work. The legs should be gently tapered to very fine needle points.

FIG. 4.—Dividers.

Use.—This instrument is used for transferring distances or spaces from one part of the drawing to another. It is also used for subdividing lines, circles or arcs. If the divider is provided with a hair-spring leg, the space to be measured is obtained by setting the instrument nearly right and making the exact adjustment by means of the hair spring.

5. Bow-Spring Instruments. (Fig. 5.)

There are usually three in a set; namely, the bow-pencil, the bow-pen, and the bow-dividers. They should be from 3 to 3–1/2 inches long and may be fitted with either wood, bone, or metal handles.

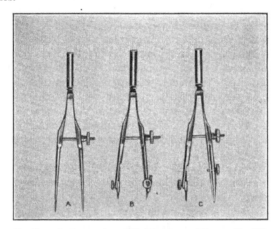

Fig. 5.—Bow instruments. (A) Dividers; (B) pencil; (C) pen.

Use.—These are used in much the same way as the larger instruments. They will, however, be found to be more convenient for drawing small circles or arcs, and for spacing. They possess the advantage, that, when set to a dimension, they are not likely to get out of adjustment.

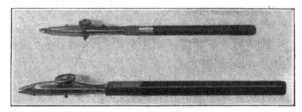

Fig. 6.—Ruling pens.

6. Ruling Pen. (Fig. 6.)

The ruling pen, frequently called a straight-line pen, should be not less than 5, nor more than 6 inches long. It may have either a bone or hardwood handle. Some are fitted with metal handles, but they are somewhat heavy and in consequence are not so easily manipulated as those fitted with lighter handles.

Use.—The pen is used to draw straight lines along the T-square blade or triangles, also curved lines by aid of the irregular curves. Ink is inserted between the nibs or blades by means of a quill which is provided on the cork of the ink bottle. The pen should never be dipped into the ink.

7. Drawing Board. (See Frontispiece.)

Care should be taken to secure a board which is made of thoroughly seasoned white-pine wood, with two cleats screwed on the under side to prevent warping. The top side should be smooth and one end must be perfectly straight and true. The edges of the board need not be at perfect right angles with each other, as only one end is necessarily used in contact with the T-square. It is an advantage, however, to have two edges trued to a right angle.

The best boards furnished by the trade have saw cuts about 2 inches apart on the under side, which in a large measure prevents the tendency to warp.

8. T-Square. (See Frontispiece.)

The blade should be of maple or pear wood, with or without transparent celluloid, or ebony edging. The head may be of any hard wood and should be without a swivel. The blade should be secured to the head by means of glue and a number of screws to insure perfect rigidity. The upper or working edge of the blade must be perfectly straight and without nicks.

Use.—The T-square is used for drawing *horizontal lines*.

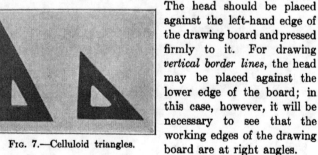

The head should be placed against the left-hand edge of the drawing board and pressed firmly to it. For drawing *vertical border lines*, the head may be placed against the lower edge of the board; in this case, however, it will be necessary to see that the working edges of the drawing board are at right angles.

Fig. 7.—Celluloid triangles.

9. Triangles. (Fig. 7.)

These may be of wood, hard rubber or celluloid. The last is preferable on account of its transparency and cleanliness. Celluloid, like wood or rubber, is likely to warp in the course of time, and triangles made from it may have to be *trued up*, which is very easily accomplished with the aid of a file or plane. The most generally useful triangles are those of 30° and 60° about 9 inches long, and those of 45° and 90° about 7 inches long.

Use.—These instruments are used in drawing lines of such angles as 90°, 75°, 45°, 30° and 15° with a given, usually, horizontal line; the triangles should rest against the blade of the T-square when in position for drawing horizontal lines.

10. Irregular Curves.

For illustrations see Frontispiece, also Figs. 49(*A*) and 49(*B*). These are also called French curves. They may be obtained in an almost endless variety of shapes and in various materials, such as paper, wood, hard rubber, or transparent celluloid. The celluloid curves are preferable on account of their transparency.

Use.—These instruments are used for drawing curves which can not ordinarily be drawn with the compass.

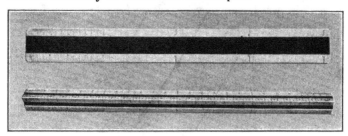

Fig. 8.—Triangular and flat scales.

11. Scales. (Fig. 8.)

A scale graduated into inches and subdivided into tenths, fiftieths, and hundredths is called an engineers' scale. It may be made of the same materials as stated in the scale described below. This scale is used generally by civil engineers in drawing maps, surveys, etc. It is not employed by the mechanical draftsman, except in unusual cases.

The triangular architects' scale, made of plain boxwood or boxwood with white celluloid edges, is used by mechanical draftsmen and contains eleven different scales. One edge is graduated into inches and subdivided into sixteenths; this is used for making full-size drawings. It may also be used for making half-size drawings by mentally dividing the dimensions by two. The other graduations are reduced scales such as: 3 inches equals 1 foot, 1-1/2 inches equals 1 foot, etc. It is obviously impossible to make all drawings full-size, therefore one or the other of the reduced scales must be employed. A number of the various graduations on the scale are as follows:

Designation	Meaning	
Full size................................	12″	= 1 foot
Quarter size...........................	3″	= 1 foot
Eighth size.............................	1-1/2″	= 1 foot
Twelfth size...........................	1″	= 1 foot

Mechanical drawings are invariably made to one of the four scales given. The more reduced scales are used for architectural drawing.

The principal disadvantage of a triangular scale are its numerous graduations, which are unnecessary and confusing to the mechanical draftsman.

The most suitable scale for mechanical drawing is a flat scale made of boxwood, with beveled edges of white celluloid, and graduated into full, half, quarter and eighth sizes.

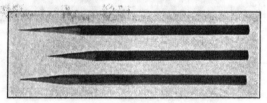

FIG. 9.—Pencils—showing long tapering points.

12. Pencils. (Fig. 9.)

The draftsman should have three pencils. One should be a 3H with a slender round point; the other two should be either 4H and 5H, or 5H and 6H, with chisel points.

Use.—The pencil with the round point is to be used for lettering, figures, and arrowheads. The harder of the other two should be used for laying out the drawing and making those lines whose limitations are not determined. The third pencil is intended for all other work in finishing the drawing.

13. Pencil Pointer. (Fig. 10.)

This consists of a number of small strips of very fine sand paper, fastened together at the edges in the form of a pad, which in turn is glued to a thin piece of wood which serves as a holder.

Use.—It is used to sharpen the lead of a pencil after the wood portion has been removed with a knife. A fine flat file also makes an excellent pencil pointer.

Fig. 10.—Pencil pointer—sand paper.

14. Erasers. (Fig. 11.)

These are made of rubber, and for pencil erasures the rubber should be soft and very pliable.

The ink eraser is also made of rubber, but has a quantity of very fine sand or pumice-stone incorporated into it during the process of manufacture. This sand or pumice imparts to it a grinding action when used. It should be hard and yet pliable.

The rubber eraser, like other objects made of rubber, will deteriorate in time, becoming hard and brittle. When it will not make erasures without scratching the paper, it should be discarded.

Fig. 11.—Pencil and ink erasers.   Fig. 12.—Sponge rubber.

15. Sponge Rubber. (Fig. 12.)

The sponge rubber is used for cleaning the surface of a drawing, which naturally becomes more or less soiled when working on it for some time. It is generally used after the inking is finished, although it may be used to a limited extent on a pencil drawing; great care, however, must be exercised not to obliterate the lines. These rubbers should be perfectly clean, very soft and pliable. The cleaning may also be done with a substance known as art gum, instead of the sponge rubber.

16. Lettering Pens. (Fig. 13.)

Nearly all lettering and figures used in mechanical drawing are made free hand and not with the aid of instruments other than writing pens. It is very desirable that letters and figures should consist of sharp, bold, clean lines and corners, of uniform strength; therefore the pens used for such work must be selected with great care and judgment. For the ordinary single-stroke *title letters*, a

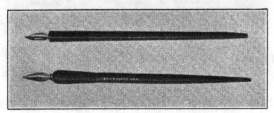

FIG. 13.—Lettering pens.

ball-pointed pen, such as "Leonardt's No. 516 F," will be found suitable, and for small letters, figures, and arrowheads, the "Soennecken, No. 108," will be found very good. For very small work, the so-called crow-quill pen is most generally used.

17. Penholders. (Fig. 13.)

The penholder should be of a size easily controlled by the fingers without cramping. Holders with cork or rubber ends will be found very satisfactory. Crow-quill pens usually come with holders of very small diameter; these will be found to be unsuitable for good work, as they can not be satisfactorily guided; the pen should, therefore, be fitted into a larger holder.

FIG. 14.—Thumb tacks.

18. Thumb Tacks. (Fig. 14.)

These are small round discs of steel, brass, or German silver with a small pointed pin in the center. The disc is slightly crowning on top and the edge is very thin, which permits the T-square to slide over it readily. The most desirable thumb tack is one in which the disc and tack are made from one piece.

Use.—They are used to fasten drawing paper and tracing cloth to the drawing board.

19. Ink. (Fig. 15.)

Previous to the introduction of prepared ink, it was customary for draftsmen to prepare their drawing ink by grinding India or Chinese stick ink, with a small quantity of water, in a saucer of slate or ground glass having a roughened surface. While this process produces a superior drawing ink, the time consumed in its preparation is considerable. Prepared inks of various colors, in bottles and of excellent quality, can now be had in the market. These prepared inks flow more freely than those produced from sticks, and can be diluted with a little water if found to flow sluggishly.

Ink produced from India or Chinese sticks can be made water-proof by adding a very small amount of bichromate of potassium.

FIG. 15.—Liquid drawing ink.

Black ink is universally used in mechanical drawing. Sometimes red or blue ink is used for center and dimension lines. While these colors are allowable to a certain extent on paper drawings, they are not to be recommended for tracings, because of their transparency. In a blue print, slightly overexposed, colored lines are almost entirely obliterated.

20. Soapstone. (Fig. 16.)

This is the mineral "Steatite" commonly known as soapstone or talc.

Use.—Mechanical draftsmen use this occasionally for re-surfacing tracing cloth where an erasure has been made, thereby enabling them to draw new lines over the surface without danger of having the ink spread. It may be used in the form of a powder and applied with a piece of cloth; or if in pencil

FIG. 16.—Soapstone pencil. form, which is more convenient, may be carefully rubbed over the surface where an erasure has been made.

21. Chamois.

Chamois skin was originally made from the hide of a small Alpine antelope called chamois, but science and chemistry now produce what is frequently sold for chamois, a soft leather made from sheepskin. This leather is satisfactory for nearly all commercial purposes, and its cost is very much below that of real chamois skin.

Use.—A small piece of chamois, or soft linen cloth free from lint, should always be used to clean and wipe drawing pens and instruments after using. Great care must be exercised to wipe the steel parts dry before laying instruments away. This care will prevent them from rusting and corroding.

PAPER AND TRACING CLOTH

22. Drawing Paper.

There are many grades of paper used for drawing purposes. In general, any paper which is tough and strong, which will not repel or absorb ink, and on which a good erasure can be made will be suitable for the general work of the draftsman.

For drawings of which tracings are to be made, a cheap grade of Manila paper will answer, as it will take good clean pencil lines. For a good drawing which is to be inked and made permanent, a better grade of paper should be used. For very fine work in ink, the paper known as Whatman's Hot Pressed is extensively used, while for drawings which are to be tinted with water color, Whatman's Cold Pressed finds favor among draftsmen.

23. The size and quality of paper must be determined by the kind and size of drawings to be made. It is supplied by the trade in sheets or rolls. Paper in rolls may be had in various widths from 30 to 72 inches and of almost any length.

The dimensions of sheets for average work are:

Cap	13″ × 17″
Demy	15″ × 20″
Medium	17″ × 22″
Royal	19″ × 24″
Imperial	22″ × 30″
Double Elephant	27″ × 40″

The size of the paper should generally be selected so that the waste will be a minimum. A good size for school use is 11 × 15 inches which is a quarter sheet of "Imperial" size.

24. Tracing Paper.

This is a thin, tough paper, specially treated to make it transparent. It may be obtained in white or light tan color; both kinds are used by draftsmen.

Use.—It is used to make copies in ink, generally from pencil drawings. This copy being transparent, is used to print from in the same way as a photographic negative, thereby producing exact reproductions of the original, in the form of blue or brown prints.

25. Tracing Cloth.

Tracing cloth is a very thin, fine, linen cloth, treated by a patented process on one side, so as to render the material both transparent and smooth.

Use.—It is used in the same way as tracing paper, but being cloth is, of course, very much more durable. Its cost as compared with that of the paper is very high. This, however, does not prevent its being universally used for tracings on account of its durability.

LIST OF SUPPLEMENTARY INSTRUMENTS

26. Rule. (Fig. 17.)

For all full-size drawings the ordinary 12-inch rule, graduated in inches and subdivided into sixteenths, will be found very

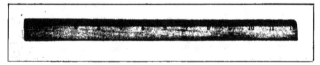

Fig. 17.—Common 12-inch rule.

convenient. These rules may be obtained in steel, plain boxwood, or boxwood with white celluloid edges. The latter will prove the most satisfactory on account of the clearness of their graduations and their general cleanliness.

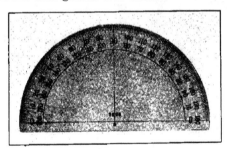

Fig. 18.—Celluloid protractor.

27. Protractor. (Fig. 18.)

A protractor is usually a semicircular disc of horn, celluloid, or other material. Those made of celluloid are desirable on account of their transparency. The outer edge is divided into degrees.

Use.—It is used for laying off and measuring degrees of angles, other than those for which the triangles can be used. The center is indicated on the straight edge of the instrument.

FIG. 19.—Pencil leads.

28. Leads. (Fig. 19.)

These may be obtained in assorted grades of hardness in one box, or may be purchased by the piece. The 4H or 5H grade is most generally used.

Use.—This lead is used in the pencil leg of the compass and bow pencil. It is also used in what is known as the tubular pencil holder.

29. Erasing Shield. (Fig. 20.)

This instrument consists of a thin plate of some such material as brass, steel, celluloid, or paper. It has various shaped openings or holes cut through.

Use.—One or the other of these openings is to be placed over the spot on the drawing where an erasure is to be made, thereby enabling one to confine the erasure to a limited area.

FIG. 20.—Erasing shield.

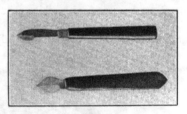

FIG. 21.—Erasing knives.

30. Erasing Knife. (Fig. 21.)

The steel erasing knife is used exclusively for erasing ink lines from tracing paper or cloth. A well-sharpened pocket knife will answer almost equally well.

Use.—To erase an ink line from a drawing or tracing, the erasing knife should be used lightly until the line has almost disappeared, then the rubber ink eraser should be used, and lastly the pencil eraser. With care and judgment this process will give good results. If the erasure is made on tracing cloth, it will be desirable to resurface the spot with a soapstone pencil, and polish with a piece of cloth. The surface thus treated will take ink without danger of blotting.

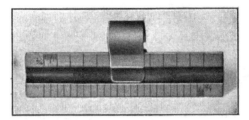

FIG. 22.—Triangular scale guard.

31. Triangular Scale Guard. (Fig. 22.)

This is usually made of brass, nickel-plated, and is used to spring over one edge of the triangular scale, thereby enabling the draftsman to readily find the scale which he desires to use. Without this guard there is difficulty in finding the desired scale, as there are five surfaces, each bearing two different scales, and reading from alternate ends.

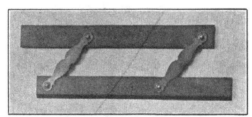

FIG. 23.—Parallel rule.

32. Parallel Rule. (Fig. 23.)

This consists of two rulers made of either wood, metal or celluloid, united by two bars. The bars are so adjusted to the rules that when the instrument is opened, the bars and rules form a parallelogram. This arrangement keeps the one rule parallel to the other.

Use.—It is used only to draw a series of parallel lines such as are required in cross-section work, but as it can not be depended upon for accuracy it is generally replaced by the T-square and triangle, or section liner.

33. Horn Centers. (Fig. 24.)

These are discs about 1/2 inch in diameter, made from some very thin transparent material such as horn or celluloid. They are provided with three very fine steel pins about 1/32 inch long, on one side of the disc.

Use.—These discs are used to fix over a given point on a

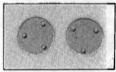

drawing, from which a number of concentric circles are to be drawn. The object of the disc is to prevent the needle of the compass from puncturing the paper, and, by continued use, making an unsightly hole. This, however, can be avoided by a careful workman

FIG. 24.—Horn centers.

without the use of a horn center.

34. Beam Compass. (Fig. 25.)

The instrument may consist of a wooden bar to which are fitted two sliding boxes for holding the needle, pencil or pen points; or it may be made with either circular or square sectional bars of metal. The latter, however, are very costly. Some of the better instruments, with metal bars, have graduations along the bar which enable one to make a setting without

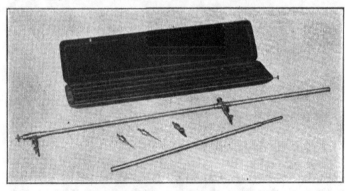

FIG. 25.—Tubular-bar beam compass.

the aid of a scale. One of the boxes usually has what is known as a fine adjustment device.

Use.—It is used for drawing arcs, circles, and transferring distances, which are too large for the ordinary compass or dividers. On account of its cost, this instrument is seldom owned by the draftsman, and where drawings are to be made which require its use, it is generally found as a part of the drawing-room equipment.

35. Proportional Dividers. (Fig. 26.)

In appearance these are very much like a pair of double-ended dividers. They are provided with a movable slide which may be fastened after both ends of the instrument are adjusted to a desired ratio.

Use.—They are generally used for transferring distances from one place to another, or in drawing to a reduced or enlarged

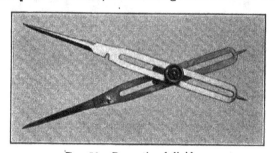

FIG. 26.—Proportional dividers.

scale. They may also be used to a certain extent for obtaining revision ratios as: diameter of a circle to its circumference; side of enclosed square to its area, also miles to knots; gallons to cubic feet, etc. A table of these various settings is furnished with each instrument by the makers. This instrument is very seldom used and should properly form part of the drawing-room equipment, rather than the individual draftsman's outfit.

36. Section Liners. (Fig. 27.)

There are a number of varieties of these instruments on the market, ranging in price from 50 cents to 20 dollars. They may be made from wood, metal or celluloid, or a combination of two or all of these materials. They are all provided with an adjustable device to vary the spaces.

Use.—Their only use is to enable the draftsman to draw a series of lines at equidistant spaces. Since cross-sectioning is generally required over small areas only, it is best to dispense with this instrument and do the work with the T-square and triangle. A little practice will enable the draftsman to produce results which meet all requirements in commercial work.

Fig. 27.—Section liner.

37. Spline and Spline Weights. (Fig. 28.)

A spline consists of a thin narrow strip of wood or celluloid about 3 feet long. It is rectangular in cross-section and has a small groove cut into one of its narrow edges. A spline weight is generally made of lead. It is used for holding the spline in position by means of a projecting "finger" on one end.

Fig. 28.—Spline and spline weights.

Use.—The spline is used principally by ship draftsmen for drawing long smooth curves. It may be used in mechanical drawing where such curves are wanted. Three or four weights are generally necessary to hold the spline in the desired position.

38. Adjustable Curve. (Fig. 29.)

This instrument is made of two bands of thin spring steel, a bar of soft lead, and a strip of hard rubber. These three materials are held together by means of small metal clamps about 1 inch apart throughout its length. The lead being a soft metal will permit its being bent into any curve and retaining its shape. The rubber is the ruling edge along which lines may be drawn.

FIG. 29.—Adjustable curve.

Use.—This instrument is used to draw irregular curves, like the curves drawn with the fixed instrument. It may be bent so that the rubber strip will coincide with a series of given points through which the line is to be drawn. It is used very little by draftsmen on account of the difficulty experienced in shaping it right, which requires time and great skill.

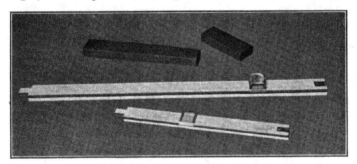

FIG. 30.—Common slide rule.

39. Slide Rule. (Fig. 30.)

The slide rule is an instrument designed for the purpose of performing various arithmetical calculations. A study of the fundamental principles which underlie its operation will enable anyone to make many calculations, after a little practice, with comparative rapidity and éase.

Slide rules may be obtained in various lengths. The size most commonly used by draftsmen and engineers is about 10 inches long, and one of good quality can be purchased for 3 dollars.

CHAPTER II

THE USE OF INSTRUMENTS

SCALE

40. Linear Measurements.

For linear measurements that scale is used which is described under the heading of instruments.

It is obvious that not all drawings can be drawn full-size. Some drawings are made to fractional proportions—one-half or quarter size, or in some cases smaller. If it is desired to make a drawing quarter size, this means that 3 inches should equal 1 foot; for this, the scale marked 3 is used. In this scale 3 inches is equal to 1 foot. (See Fig. 31.) The first distance of 3 inches is divided into twelve equal parts, each part representing 1 inch, which in

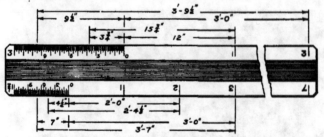

FIG. 31.—Reading the scale.

turn is subdivided into smaller divisions such as 1/2 and 1/4 inch. Inches are taken to the left of the zero indicated on the scale, while feet are read toward the right. If, for example, it is desired to lay off 3 feet 9-1/2 inches, the feet are taken to the right from the zero and the 9-1/2 inches are added to this distance from the subdivided foot, giving the full distance required. Or, if 15-3/4 inches are wanted, the foot is taken to the right, and the inches to the left, of the zero mark.

Many draftsmen set their dividers to the required measurements on the scale, then transfer this distance to the drawing.

20

While this practice gives good results, it is not to be commended, as it tends to destroy the graduations on the scale. It is better to place the scale on the drawing paper and mark the measurements by means of a fine pointed pencil or a needle point.

PROTRACTOR

41. Circular Measurements.

For laying off angles other than those which can be obtained with the triangles, or for dividing circles into any number of parts, the protractor is used. This instrument is generally divided into degrees. Protractors used by draftsmen are usually made of celluloid, graduated to half degrees, which for general work are accurate enough. More elaborate protractors, having a vernier attachment, enable the draftsman to measure angles to a fraction of a degree.

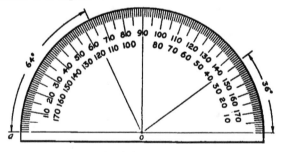

Fig. 32.—Reading the protractor.

To mark off an angle of 64° from a given line, place the radius *oa* on the given line, with *o* at the vertex of the required angle; then with a sharp pointed pencil make a dot opposite the required division. (See Fig. 32.) A line drawn from this dot to the point *o*, will complete the required angle.

To divide a circle into any number of equal parts, it is only necessary to divide 360 by the required number of parts; the quotient will give the angle required, which may be laid off by the method just described.

TRIANGLES OR SET-SQUARES

42. Non-horizontal Lines.

Two triangles are used by the draftsman for general work, one called a 45° triangle, and one a 30° and 60° triangle.

They may be of wood, hard rubber, or celluloid. The latter
is by far the most desirable material. While it is true that
celluloid triangles are likely to warp somewhat, they have the
advantage of enabling the draftsman to see what is underneath.
They do not soil the paper as much as those made of hard
rubber and can be readily cleaned with soap and water. Triangles made of thoroughly seasoned celluloid will retain their
accuracy for a considerable length of time. If, however, they do

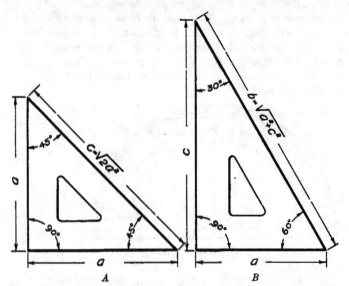

FIG. 33.—Showing the relations of the sides of triangles.   (A) 45° triangle;
(B) 30° and 60° triangle.

lose some of their *trueness*, they may readily be corrected with a
plane or fine file. For general work a 7 inch 45°, and a 9 inch 30°
and 60° triangle are most convenient. For large work larger
sizes may be obtained.

By referring to Fig. 33(A), it will be seen that a 45° triangle
has two of its sides of equal length, while the length of the third
side c, as geometry shows, is equal to the square root of twice the
square of one other side, while in the 30° and 60° triangle [Fig.
33(B)] the length of the side b is equal to the square root of the
sum of the squares of the other two sides. The length of the

side b is also twice the length of the side a. A knowledge of the properties of these triangles will prove helpful should the student desire to make triangles for himself.

LEAD PENCILS

43. Hardness of Pencils.

The degree of hardness of a pencil is usually indicated by the letter B or H prefixed by a number; thus 5B, which is the softest, to 9H, the hardest.

For detail drawing, pencils ranging from 3H to 6H may be used. A 5H or 6H should be used for drawing center lines and lines whose exact limitations are not defined. For lines whose limitations are known a number 3H or 4H may be used, while for lettering and figures a number 2H or possibly 3H will be found suitable.

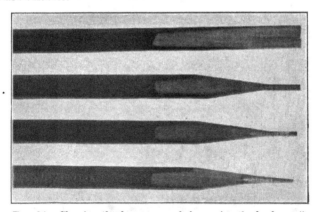

Fig. 34.—Showing the four stages of sharpening the lead pencil.

44. Pencil Points.

To obtain good results in drawing considerable attention must be given to the point of the pencil; it must not be allowed to become blunt or rough by the breaking of the point. There are two kinds of points in common use—the round or conical point, and the flat or chisel point.

A nicely rounded point should always be used for lettering and dimensioning. Some draftsmen use the round point for all ordinary drawing, although this point will wear more quickly than the chisel point; two or three lines drawn with the round point will dull it so that repointing is necessary. The chisel-pointed end finds favor with many draftsmen because it will longer retain a good working edge.

45. Sharpening Pencils.

To sharpen a pencil for a round point, begin about 1–1/4 inches

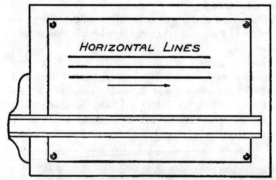

FIG. 35.—Showing horizontal lines.

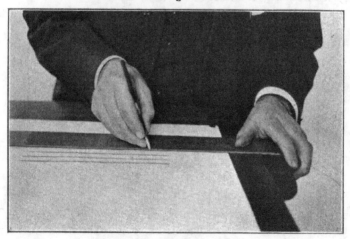

FIG. 36.—Drawing horizontal lines.

from the end and cut away the wood to form a true cone, leaving the lead projecting about 1/4 inch; round this to a fine conical point by rubbing across a pencil pointer or a fine file, while rotating the pencil between the thumb and the index finger.

46. For the wedge-shaped or chisel point the wood is cut away on opposite sides, beginning about 1–1/2 inches from the end and tapering to a true wedge shape, with the lead projecting about 1/4 inch. The remaining four faces, if the pencil is hexagonal,

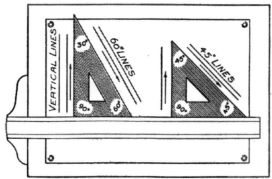

Fig. 37.—Showing vertical and oblique lines.

Fig. 38.—Drawing vertical lines.

are then cut away, from a distance of about 1 inch back to within about 3/16 inch from the end of the lead. The lead is then fashioned with the pencil pointer on the two long sloping sides to a sharp wedge, then narrowed down a trifle at the point, if desired. For the four stages of sharpening, see Fig. 34.

47. Drawing Lines.

The draftsman usually stands by a longer edge of the drawing board and that position determines the naming of the lines he draws. All lines drawn parallel to that long edge are called horizontal lines; all lines at right angles, vertical lines.

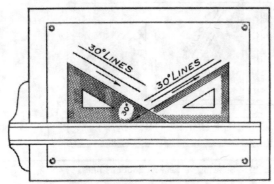

Fig. 39.—How 30° lines are drawn.

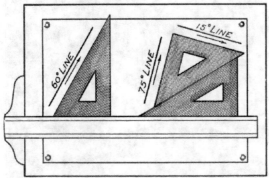

Fig. 40.—How 15° and 75° lines are drawn.

Lines should be drawn in definite directions; the proper direction in each case is shown in the various illustrations by arrows.

48. *Horizontal lines* (Fig. 35) are always drawn with the T-square. The head of the T-square is firmly pressed against the left-hand edge of the drawing board and the line is drawn along the upper edge of the blade (Fig. 36). When a long line is to be drawn, keep the blade from springing by sliding the left hand along the blade, after it is set to the proper position. By moving the T-square up or down on the drawing board, any number of parallel lines may be drawn.

49. *Vertical lines* (Fig. 37) are always at right angles to horizontal lines. They may be drawn with either of the two tri-

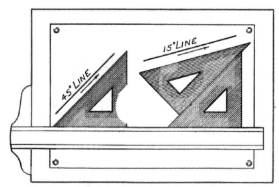

Fig. 41.—Another arrangement for drawing 15° lines.

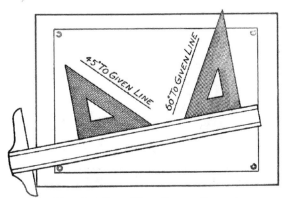

Fig. 42.—How oblique lines are drawn.

angles by placing one edge of the triangle against the T-square, as shown in Fig. 38, and holding both the T-square blade and triangle with the left hand. The triangle is always placed as shown and the lines are drawn by moving the pencil along its edge. It is customary to draw vertical lines with an upward

movement of the pencil or pen instead of downward, the latter
stroke being a somewhat awkward movement to make when the
T-square and triangle are in proper position.

50. Lines making angles of 30° with horizontal lines may be
drawn with the 30° and 60° triangle by placing one of its edges
against the T-square blade as illustrated in Fig. 39.

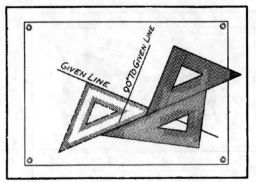

Fig. 43.—Method for drawing a line perpendicular to a given line.

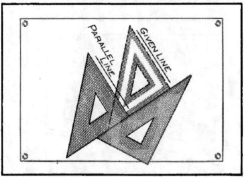

Fig. 44.—Method for drawing parallel oblique lines.

51. Lines making angles of 60° with the horizontal are drawn
with the 30° and 60° triangle by placing its short edge against
the blade as shown in Figs. 37 and 40.

52. Lines making angles of 45° with the horizontal are drawn
with the 45° set-square as illustrated in Figs. 37 and 41. The
method and direction for drawing the lines are the same as
described for the 30° lines.

53. By using both triangles and the T-square, lines making angles of 15° may be drawn. The angle may be obtained by using the combination of triangles as shown in Fig. 40, or by reversing the order as in Fig. 41. By sliding the upper triangle along the edge of the lower, any number of parallel lines may be drawn, each making an angle of 15° with a horizontal line.

54. If it is desired to draw lines, at an angle to a given line, in any direction other than horizontal, the method illustrated in Fig. 42 may be used. In this case the blade of the T-square is placed against the given line and the proper triangle placed against the blade. If the given line is short, use one of the triangles instead of the T-square blade, as the edge against which to place the other triangle; then draw the required line. Any number of parallel lines may be drawn as the parallel lines of 15° are drawn.

55. To draw one or more lines at right angles, or parallel to a given line, such an arrangement as shown in Figs. 43 or 44 may be used. Place one triangle against the given line and one edge of the other triangle against the first. If the 45° triangle, as illustrated, is moved along the edge of the 30° and 60° triangle, any number of lines parallel or perpendicular to the given line may be drawn. In the figures but one line is shown.

After some experience the student will have but little difficulty in selecting the most convenient arrangement of triangles and T-square to draw lines parallel to a given line; or lines making any required angles with given lines, and in any direction.

RULING PEN

56. Care of Ruling Pen.

The Ruling Pen is one of the most important instruments which a draftsman uses. On its skillful use depends the good appearance of a drawing. To produce good results the pen must always be in prime condition. As it is in constant use it requires frequent sharpening; a pen which is dull or which has uneven length of blades will invariably produce lines ragged at the edges and not uniform in width; fine lines can not be drawn with a *dull* or *untrue* pen.

57. Sharpening the Ruling Pen.

To sharpen the pen, screw the blades together until they touch. Hold the pen over a piece of white paper; if the ends are uneven the need for sharpening is indicated. To sharpen, place the pen in a perpendicular position on a fine-grained oil-stone and move it back and forth in a rocking motion changing to an inclination of 30° on one side of the stone to 30° on the other, thereby producing an elliptical end. Examine frequently and stop when the ends are even and nicely rounded. A small magnifying glass will be of great help when examining the ends. After very little rubbing the blades will be of the same length but very blunt. Then open the blades to about one-eighth of an inch and sharpen on the *outside*. The *insides* of the blades must not be touched with the stone or they will become convex or rounded, which destroys the usefulness of the pen. The outside must be sharpened until the bright spot at the end, produced by the first operation, disappears, at which instant the rubbing should cease. The ends must not be sharpened to a cutting edge, or they will cut into the paper or tracing cloth. If they are found to be too sharp, they can easily be dulled sufficiently by drawing a number of lines on a piece of waste paper.

58. Placing Ink in the Pen.

If prepared bottle ink is used it should be shaken to restore it to a uniform density.

Hold the bottle down with the second and third fingers of the left hand, and with the thumb and index finger of the same hand remove the cork and touch the quill between the blades of the pen, which should be held in an inclined position with the blades down. Do not perform this operation *over* the drawing. After the pen is inked, test it on a separate piece of paper. If it will not draw *freely*, touch the end of the blades with the tip of a moistened finger; if it still refuses, clean the pen thoroughly with a piece of cloth or chamois and refill. A pen which is not free from dried ink and otherwise thoroughly clean will not give good results. The pen should be wiped clean after each charge of ink has been exhausted.

After filling the pen, be sure that no ink has lodged on the outside of the blades. If this precaution is not taken, smudging or blotting may occur.

59. Inking Straight Lines.

Hold the pen in the right hand and with the second or third finger and thumb adjust the screw; test on a piece of trial paper until the desired width of line is found. Place the pen against the T-square blade or triangle, with its blades parallel to the direction in which the line is to be drawn. (Figs. 36 and 38.) Hold the pen at the starting-point for just an instant until the ink begins to flow, then move it with a moderate speed toward the right, if along the T-square; or toward the top, if along the triangle. On reaching the end of the line the pen should immediately be raised to prevent the ink from being drawn out by the paper, which will act more or less like a blotter. It is well not to load the pen with too much ink, for fear of blotting.

Do not bear down on the pen and do not press it too hard against the T-square blade or triangle, as this would have a tendency to close it, thereby producing a line of varying width. To obtain a good line, the motion of the pen should be smooth and uniform from the beginning to the end of the line. If the pen is moved forward too rapidly, a very irregular line will be the result.

When a number of lines are to meet at a corner or point, it is well, if possible, to draw *from* the point and not *toward* it, thus avoiding an accumulation of ink which may cause a blot. Be sure to allow a sufficient amount of time for one line to dry before the next is drawn.

Fig. 27 shows a section liner in position for drawing a number of parallel lines, such as are used on cross-sections.

COMPASS

60. Circles and Arcs.

Circles and arcs of circles are drawn with the compass, either in pencil or ink. A complete compass consists of five parts:

The Compass (proper).
The Pencil Leg.
The Pen Leg.
The Extension Bar.
The Needle Point.

For use: The needle point should be slightly (very slightly) longer than the pencil or pen leg. This point should be adjusted to the pen leg, and the lead of the pencil leg adjusted to the point. The pen and pencil legs are then interchangeable without the necessity of altering the length of the needle point. For drawing small circles, up to about 4 inches in diameter, the legs of the compass may be straight. To draw large circles, bend the hinges so that the legs will be about parallel to each other, and perpendicular to the paper. (See Fig. 45.) To draw a circle, place the needle point in position using it as a pivot,

FIG. 45.—Beginning a circle.

FIG. 46.—Completing a circle.

slightly incline the instrument and, with a little pressure exerted at the head, turn the compass about as the hands of a clock move.

Fig. 45 shows the beginning of a circle to be drawn, while Fig. 46 shows the circle nearing its completion.

NOTE.—The steel pin or attachment, designated as the needle point, usually has a long conical point at one end and a sharp thin needle point at the other end. The conical point is intended for large drawings and detail work, the needle point for fine work.

61. The instrument should be manipulated with the right hand only. The opening or closing of it for the purpose of changing the sweep and drawing a line can be very easily accomplished with one hand, after very little practice. If the compass is adjusted for a large radius or sweep, the left hand may assist by guiding the needle point to its proper place.

62. When the circle is to be drawn in ink, the pencil leg must be removed and the pen leg inserted. The pen is inked in the same way as the ruling pen.

For circles of larger radius, the extension bar is inserted, after removing the pencil or pen leg, and either pencil or pen, in turn, inserted in the lower end of the bar.

For very large circles it may be desirable to guide the marking leg with the left hand to keep it steady.

DIVIDERS

63. Dividing or Transferring Distances.

The dividers have two fixed legs both of which are tapered to needle points. Great care should be taken to guard these points from injury. Some instruments are provided with a hairspring adjustment for a very fine degree of accuracy in setting. It is used for transferring distances from one part of the drawing to another. Many draftsmen use them for transferring measurements from the scale to the drawing, but this injures the graduations on the scale.

64. The principal use for the dividers is to space or divide a given distance into a definite number of equal parts by successive approximations. This is accomplished by setting the legs to approximately the correct distance, setting off the spaces on the line to be divided, by rotating the instrument between the thumb and fingers in the same manner as the compass is handled, revolving one point over the other in a direction from left to right. (See Figs. 47 and 48.) If the spacing so made is not correct, the legs are readjusted and another trial made. With a little practice the desired divisions can be made with very few trials. In the same manner the circumference of a circle may be divided into any number of equal spaces,

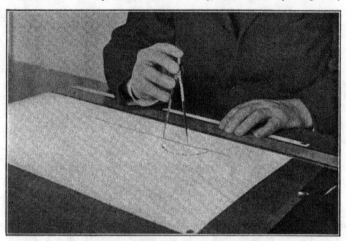

FIG. 47.—Using the dividers.

FIG. 48.—Dividing a line by trial.

or the periphery of a circle measured. In the trial spacing press lightly so that the surface of the paper may not be perforated with unsightly holes.

IRREGULAR CURVES

65. Non-circular Arcs.

These instruments are designed for drawing curved lines which are not circular arcs; they may, however, be used for short circular arcs whose radii are too large for the ordinary compass, or when a beam compass is not available.

66. To ink an irregular curve, passing through a number of fixed points and producing a line which will be smooth and pleasing to the eye, is one of the most difficult operations the draftsman has to perform, and nothing short of persistent effort and continued practice will enable him to produce good results.

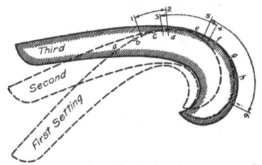

FIG. 49(*A*).—Using the irregular curve.

The proper method is to sketch a free-hand line as smoothly as possible through the given points. Then apply to this line that section of the curve which will best coincide with the greatest or longest portion of the pencil line drawn. In inking any part of this completed pencil line, care must be taken to turn the pen so that the blades will at all times be tangent to the curve, otherwise the line will be of varying width. In inking the line, it is advisable not to ink the full length of the line *matched* by the curve, but to move the curve to a new position and resume the inking. By referring to Fig. 49(*A*), it will be seen that the curve in the first setting passes through points *b*, *c*, and *d*. The inking of the line, however, should extend from somewhere near points 1 and 2. The curve is then shifted to the second position and the next section is inked, which extends from points 3 and 4; then the section which lies between points 5 and 6 is inked, making a smooth curve which begins at *a* and terminates at *i*.

In many cases it may be desirable not to join the adjacent sections, but to leave a small opening which may be filled in with a separate setting of the curve. By this method, while the line must of necessity be composed of more sections, and more settings of the curve are necessary, the additional time required may very often be well spent.

67. When a curved line is symmetrical about an axis, as is the case with conic sections or other curves having sharp turns, the beginner invariably has great difficulty in avoiding a break, or a more or less sharp point at the turn. This difficulty may be overcome if the line is inked a trifle short of the sharpest turn of the curve, and the remainder of the curve drawn either with a new setting, or with the compass.

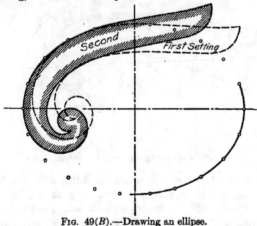

Fig. 49(*B*).—Drawing an ellipse.

Fig. 49(*B*) shows the location of a series of points for an ellipse. This is a case where a combination of the compass and irregular curve may well be employed.

Very often, if an especially good curved line is wanted, it may be desirable to make a "template." Any student with a little mechanical ability can make a template for some special curved line. It may be from a thin piece of wood or even cardboard, but a thin piece of sheet celluloid makes the very best. Cut the sheet with a pair of scissors and trim it to the line with a fine file or a knife.

An irregular curved line may also be drawn by joining a number of circular arcs, which pass through the given points. This method will in many instances give better results than using the irregular curve.

CHAPTER III

METHOD OF PROCEDURE IN DRAWING

68. Preparation of the Paper for Work.

Exercises and problems should be executed on sheets of paper measuring 11 inches by 15 inches. To fasten the paper, place the sheet about 2 inches from the left-hand end of the board and as far from the bottom as convenient, say 3 or 4 inches, if the width of board will permit. Now place the T-square on the paper and adjust its upper edge to the blade of the square, force a thumb tack in the upper left-hand corner and stretch the paper toward the right by moving the left hand along its edge, and place a tack at the upper right-hand corner; stretch the paper toward its lower edge and tack its corners. All thumb tacks must be pressed down so they will be in close contact with the paper.

69. Marking off Measurements.

In marking off a series of measurements along a given edge or line, it will be well to place the scale in position, with its zero mark to the point from which the measurements are to be taken, and lay off the distances without moving the scale. This can be done best by laying off the first distance, then adding the second required space to the first, and marking it, and so on for the succeeding spaces. This method avoids the possibility of accumulative errors and so leads to more accurate results than can be had if the scale is moved for each measurement.

70. Measurements for the Border Line.

All drawings should have a border line which will measure 9 inches in the vertical and 12 inches in the horizontal directions. There should be a margin of 3/4 inch at the lower, upper and right-hand edges, and 1-1/2 inches at the left-hand edge. (See Figs. 50 and 51.) The wider margin at the left will permit binding the drawings into book form if desired.

71. Drawing the Border Line.

Place the scale along or parallel with the left-hand vertical edge of the paper, and mark off with the round-pointed pencil, beginning at the bottom, the following distances in inches: 1/4, 3/4, 9, 3/4 (Fig. 50). Through these points draw horizontal lines with the aid of the T-square. Next place the scale along the lower horizontal edge and mark off, in inches, beginning at the left, 1/4, 1–1/2, 12, 3/4 (Fig. 51). Through these points draw vertical lines, using the triangle in connection with the T-square,

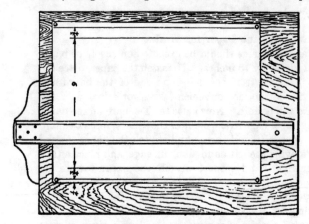

Fig. 50.—Drawing horizontal border lines.

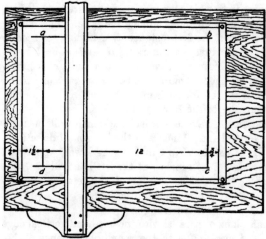

Fig. 51.—Drawing vertical border lines.

or if the triangle is not long enough, the T-square may be used. *The use of the T-square for drawing vertical lines should under no circumstances be resorted to except when drawing border lines.*

72. If the lines are drawn according to directions, there will be a small margin.   Trim this off when the drawing is completed, cutting off the strips having the thumb-tack holes, thereby making all the drawings to the uniform size which is 14–1/4 × 10–1/2 inches.

73. Sequence for Penciling.

All work which is to be successful should show skill and possess character and style; such work must proceed with system and be done methodically.   No rigid system of rules for penciling can be laid down because of the great variety in drawings; nevertheless, general directions for good practice can be indicated as follows:

1. Draw the border and cutting lines.
2. Lay off the space required for title.
3. Decide on the number of views required.
4. Make a rough free-hand sketch in the note-book of the views decided upon.
5. Calculate as accurately as possible the amount of space each view will require.
6. Lay out with as few lines as possible, the position of each view.
7. If the arrangement of the views does not present a harmonious balance to the eye, make such changes as will effect this.
8. Draw the necessary horizontal and vertical center lines.
9. Draw the limiting horizontal outlines of all views.
10. Draw the limiting vertical outlines of all views.
11. Complete all views.
12. Draw in the dimension lines.
13. Fill in the dimensions.
14. Add explanatory notes, if required.
15. Block out the title.
16. Carefully scrutinize every part of the drawing to make sure nothing has been omitted.

74. Sequence for Inking.

For inking, a sequence is more necessary than for penciling, owing to the necessity of changing instruments and waiting for ink to dry; nevertheless a good general plan is as follows:

1. Ink all small circles and circular arcs with the bow pen.
2. Ink large circles and circular arcs with the compass.
3. Ink irregular curves.
4. Draw horizontal lines beginning with the uppermost, if possible.
5. Draw vertical lines by the aid of a triangle and T-square, beginning at the left-hand end of the drawing.
6. Draw all 30°, 60° and 45° lines.
7. Draw other oblique lines.
8. Draw horizontal and vertical center lines respectively.

(NOTE.—In some cases it may be desirable to draw center lines at the very beginning.)

9. Draw section lines on all cut surfaces.
10. Draw extension and dimension lines.
11. Put in dimensions and explanatory notes.
12. Ink in title.
13. Draw border lines.
14. Carefully examine every detail of the drawing for possible omissions.
15. Clean the drawing.
16. Trim to the cutting line.

NOTE.—When the pen or compass is set for a certain width of line, do not change it, if possible, until all lines of that width are inked.

75. Inking on Tracing Cloth.

In commercial work it is customary to apply the ink on the dull side of the cloth because it works more freely, but erasures can not be made so readily because the cloth fibers absorb the ink. The beginner in mechanical drawing should use the glossy side, which if rubbed over with a little powdered chalk before working on, will readily take ink.

76. Erasures on Drawing Paper.

Pencil lines may readily be erased by lightly rubbing the line with a soft rubber.

To make an ink erasure on paper, rub the line with the ink eraser, which should be hard but flexible, until the line is indistinct; then rub over it with the pencil eraser. If the line to be erased is quite heavy, it will be well to remove the greater part of the ink with an erasing knife, using great care not to cut into or injure the paper. This may be accomplished by lightly scratching the ink until it is nearly all removed; then use the rubber ink eraser, and finally the pencil rubber.

After an erasure has been made, the paper should be thoroughly dusted with a piece of cloth to remove all particles of rubber which might otherwise get into the path of the pen and cause a blot or smear.

77. Erasures on Tracing Cloth.

Erasures on tracing cloth are very often difficult to make without causing permanent injury to the surface. In working on tracing cloth, all possible care and precaution should be taken to avoid the necessity of making erasures. If, however, the necessity arises, an erasure may be made by first removing the top layer of ink with the knife, which must not be allowed to touch the surface of the cloth, then carefully dusting off the tracing. Again use the knife, very cautiously, and finally the pencil rubber. The erased spot may be resurfaced by polishing with a soapstone pencil, after which the surface can be inked again. When inking over an erased part, it may be well to set the pen finer and go over the line several times, letting each dry before another line is drawn.

Where large surfaces have to be erased, it is well to use finely powdered pumice. This is rubbed over the spot with the end of the finger or a piece of soft cloth, and the soapstone applied as before described.

78. Cleaning of Completed Drawings.

Drawings must be cleaned with a soft rubber, "kneaded rubber" or sponge rubber. Dry bread crumbs rubbed over the drawing paper with a piece of cloth will also produce a clean surface. In cleaning, great care should be taken not to rub the ink lines too vigorously lest they lose their blackness and brilliancy.

The surface of tracing cloth which has become soiled may be cleaned with a soft cloth moistened with gasoline.

CHAPTER IV

LETTERING

79. Importance of Lettering.

Well-made letters and figures are of great importance on a mechanical drawing. Drawings on which many hours of labor have been spent in producing a good, accurate and well-planned representation, make a very unfavorable impression if the lettering and dimensioning are poorly done. There is nothing on a drawing, indeed, that contributes more to its effect than good lettering and good figures. Few students at the beginning of a course in drawing appreciate their value; but to letter well is as essential for the draftsman as to write a good hand is for the accountant.

Good letters need not be drawn with the compass and triangle, or other instruments; simplicity in style and uniformity in effect are better than mechanical exactness. The letters should be of simple outline, all of the same slant, of good proportion, and properly spaced. If these requirements are fulfilled the results will be good.

The beginner must carefully study the proportions of letters and figures which have been evolved through continued usage, and are accepted as standard forms and proportions, both with regard to design and to spacing. (See Figs. 53 and 54.) It does not always follow that a good writer can letter well, nor that one who writes poorly can not learn to make good letters. Every one is apt to letter badly at the start. The beginner will, therefore, need to practice considerably in drawing lines and curves of proper slant and of uniform spacing. With such practice, the draftsman will not follow rules committed to memory, but will depend upon his eye for judging proportions, slant, and spacing.

80. Style of Letters.

The most important style of letters used by the draftsman is known as the "Single Stroke Gothic," which is very simple in outline and comparatively easily drawn. This style of letter will be acceptable in almost any drawing and is therefore the only one which is explained and of which examples are given. It is used on all the plates in Chapter VII.

42

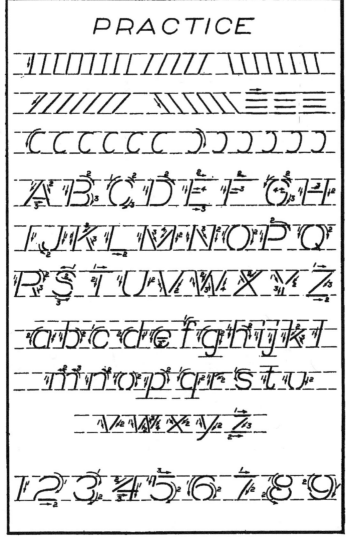

Fig. 52.—Order and direction of strokes.

FIG. 53.—Gothic letters and figures.

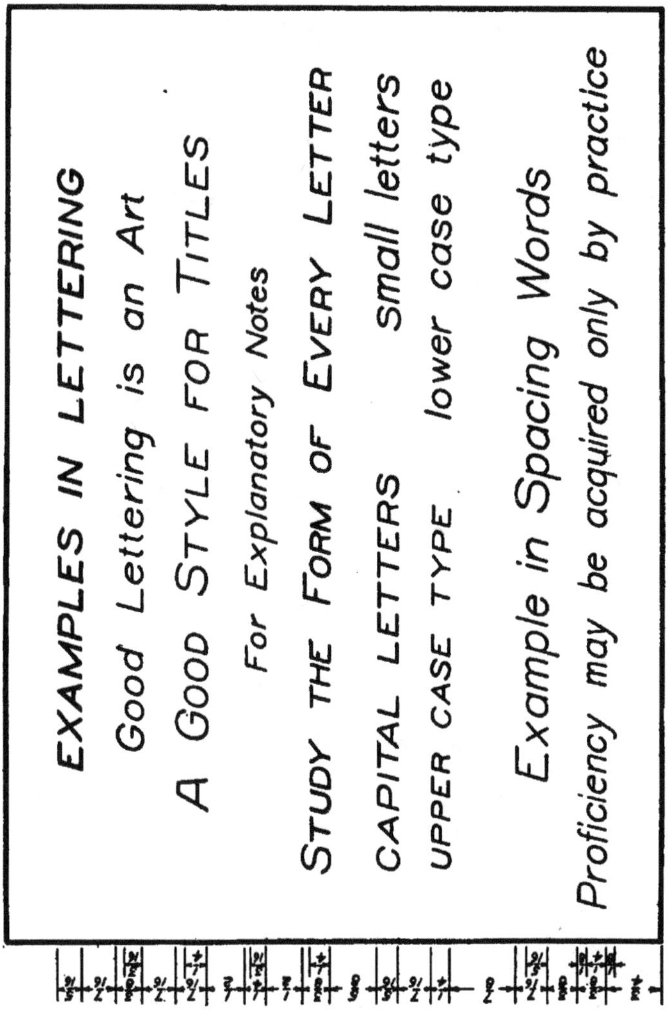

FIG. 54.—Examples of word spacing.

81. Height of Letters.

Those proficient in lettering, draw the rounded or pointed ends of letters a trifle above or below horizontal limiting or guide lines, to offset the optical illusion of having them appear shorter than those which have straight ends.

The beginner in lettering will invariably, if he attempts this, make the excess in height too great, thereby producing an effect which is more objectionable to the eye than letters made all of the same height. Since the making of letters is a question of design, it will be well for the beginner to confine their height to guide lines, and make extensions above or below these lines after he has had some experience. He can then better judge the amount of extension necessary to produce the proper effect.

82. Width of Letters.

In studying the figures the student will observe that letters vary considerably in width. If the letter H is taken as a standard for width, such letters as A, V, Y, etc., must be made wider to produce the proper effect in a word. Rounded letters such as O and Q are also made wider than those letters which have two parallel sides. The letter W is the widest in the alphabet while J is the narrowest, except the letter I.

83. Angularity of Letters.

There is no generally accepted standard of angularity or slope for inclined letters. The inclination may be any angle from 60° to 75°. The angle used in the illustrations and examples in this book is 75°. This angle can readily be obtained with the triangles as illustrated in Fig. 40.

In practice both the vertical and inclined letters are used, but the latter are preferred on account of the greater ease with which a *free-hand* inclined line can be drawn.. If a vertical line is not exactly true, the irregularity is more easily detected by the eye than is a small deviation from an inclined line.

84. Spacing of Letters.

The student will observe by studying the examples given, that the amount of space between letters will vary according to the combination required in forming a word. This space is quite as important in producing good results as the correct formation of the letters themselves. If the spacing is faulty, a word will look awkward. The proper amount of space between the letters is largely a matter of judgment and artistic sense. In forming a word it will be well to pencil one letter at a time, and after each letter estimate the proper interval of space before drawing the next. It is not necessary that each letter should be completed in its entirety before the next is begun, but enough of it should be drawn to enable the student to estimate the space necessary to give the word a proper "balance." When the word is completed it should be subjected to careful scrutiny, and correction or alteration made, if necessary, before the next word is begun.

85. Guide Lines.

As an aid to the beginner it will be well in addition to horizontal guide lines to use inclined guide lines. These lines will serve as a guide to the eye in judging the proper slope. After some experience the use of inclined guide lines may be omitted, although horizontal guide lines are always used, even by the most experienced, for all lettering, whether for titles or for explanatory notes.

86. Penciling Letters.

First the horizontal limiting lines are drawn in pencil, and if the letters are large a number of intermediate lines are drawn. In general, only the two limiting lines are necessary for letters up to 1/4 inch in height, although a third line drawn midway between these two may be found helpful in locating such horizontal lines in letters, as lie between the limiting lines. The second step will be to mark off, on the lower line, the widths of the letters and the spaces. The inclined guide lines are then drawn, after which each letter should be carefully penciled before the ink is applied.

87. Pens for Lettering.

For large letters, as in titles, a ball-pointed pen known as Leonardt's Number 516 F is used. This pen will be found very satisfactory for letters from 3/16 to 5/16 inch high. For letters from 1/8 to 3/16 inch high Soennecken's Number 108 pen will be suitable; while for letters under 1/8 inch high and for figures and arrow points, Gillott's Number 303 will give good results.

88. Inking Letters.

As was explained before, the student should first carefully study the form of each letter and the direction in which the strokes are to be made. The proper direction of strokes is of vital importance when inking, and a careful study of them is necessary to produce good results. It will be found helpful, and the time will be well spent, if the student would make a plate consisting entirely of practice strokes, such as are shown at the top of Fig. 52, before beginning a plate consisting of letters or words, as in Fig. 53. When inking, first place the pen at the beginning of the line, if the letters are drawn in pencil, and with a little pressure press the nibs apart to the desired width, hold it for an instant, then move it in the direction of the line with a uniform motion to the end; again hold it for an instant, then raise it from the paper. If the strokes are made too rapidly a line of varying width will result; the same may be said of a line which is drawn too slowly. A suitable speed for inking, to produce lines of uniform width, will be found after some practice.

If a line drawn is too thin and weak, a little more ink should be carried in the pen. If a lump accumulates at the end of a line, it is a sign that the pen is *overloaded*. The pen should always be in prime condition, free from dried crustations of ink, and otherwise perfectly clean. While in use it should frequently be wiped with a piece of cloth free from lint, and should be thoroughly cleaned when the work is completed.

89. Practice Plates.

Fig. 53 should be accurately copied for practice. Suitable heights and distances are given on the margin of the plate, although these may be varied according to circumstances. The widths of letters and spacing for each line should be laid off on a strip of paper about 1 inch wide. This *trial strip* can then be adjusted to the plate a trifle below the lower limiting line, and the spaces transferred to the plate by the aid of the T-square and triangle, after which the letters can be formed with a 3H round-pointed pencil. The entire plate is to be penciled before ink is applied to any one part.

Fig. 54 shows a number of examples of word spacing, which may advantageously be copied for practice.

NOTE.—The dimensions for location should not appear on the completed drawing.

90. Trial Title.

Before placing a title on a drawing, it would be well to make a *trial title* on a separate piece of paper before the work is done on the drawing, as several attempts may be necessary to produce a satisfactory result. If the trial title is found satisfactory, it can then be copied on the drawing.

For a more extended treatise on lettering see "The Essentials of Lettering" by French and Meiklejohn.

CHAPTER V

DRAWING ROOM PRACTICE AND CONVENTIONS

DEFINITIONS OF DRAWINGS

91. Mechanical Drawing.

Drawings executed by the aid of instruments are called mechanical drawings. Generally the term is associated with drawings representing machinery only. But all drawings which must be accurately made, and for the execution of which recourse must be had to instruments, may be classed among mechanical drawings. Geometrical drawing, descriptive geometry, architectural drawing, constructive drawing, engineering drawing, machine design and instrumental perspective, all these are mechanical drawings. In practice, the various classes are, however, differentiated and are known by the above names.

92. Assembled Drawing.

An assembled drawing is one in which all parts entering into the construction of an object or machine are shown together with each part in its proper position. The drawing need not necessarily show all the hidden lines or even all dimensions, if it be complex. It should, however, show all the principal parts and principal dimensions.

93. Detail Drawing.

A detail drawing is one in which each unit composing an object or machine is drawn separately, with all dimensions and explanatory notes for the guidance of the workman. Several details may be drawn on one sheet, but each detail must be a unit drawing, so that each piece can be made without reference to any other detail.

94. Working Drawing.

A working drawing, also called a shop drawing, is one intended to impart definite graphic information to a workman for making or manufacturing the object represented. An assembled or a detail drawing, may be a working drawing, if it contains all the necessary information the workman requires for the construction of what is represented in the drawing.

49

ALPHABET OF LINES

95. The lines used in mechanical drawing may be called the
Alphabet of Lines or **Conventional Lines.** If all lines were

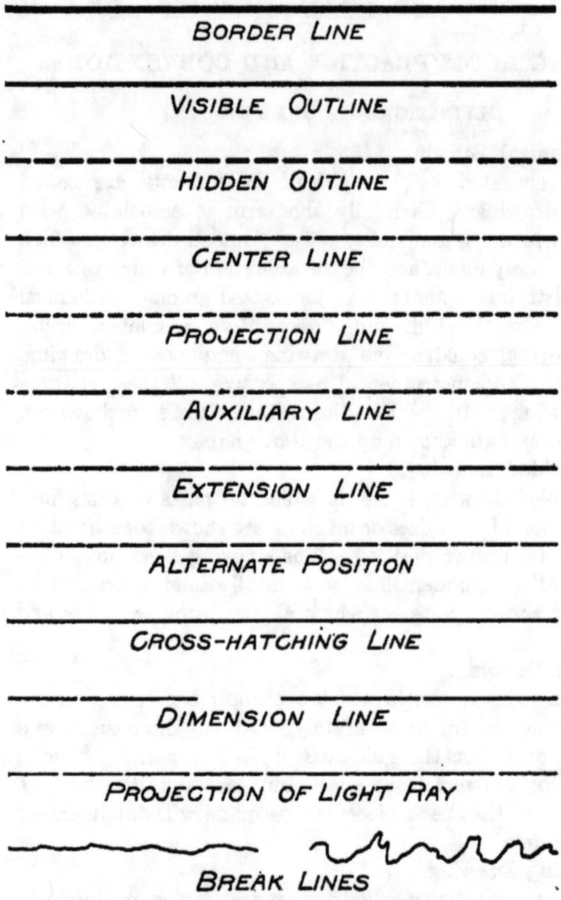

BORDER LINE

VISIBLE OUTLINE

HIDDEN OUTLINE

CENTER LINE

PROJECTION LINE

AUXILIARY LINE

EXTENSION LINE

ALTERNATE POSITION

CROSS-HATCHING LINE

DIMENSION LINE

PROJECTION OF LIGHT RAY

BREAK LINES

Fig. 55.—Alphabet or conventional lines.

identical in character, confusion would arise at times and a great
amount of time would be wasted in trying to decipher their
meaning.

In the "alphabet" illustrated in Fig. 55 there are twelve

lines. The lines are shown in about their proper width for general drawing; they may, however, be varied somewhat in width, depending upon the size of the drawing and the accuracy of detail required.

Variations in width of line may be secured by varying the pressure, or by the hardness of the pencil, if the drawing is not to be finished in ink; or, if ink is used, by opening or closing the drawing pen.

Drawings which are to be inked, are frequently drawn in pencil, making all lines continuous and of uniform density, and any desired distinction is made at the time the inking is done. This is not good practice, for a line is frequently drawn solid, in ink, when it should have been broken; then erasures become necessary, and a poor drawing results.

Beginners in drawing should cultivate the habit of drawing the pencil lines as the lines are to appear in the drawing when finished in ink. A variation in width of pencil lines is not necessary if the drawing is to be inked.

96. Border Line.

The border line is very often used to enclose a drawing as a frame, which adds to its appearance. For general purposes, however, the border line for economy of time is omitted. Sometimes a border is drawn in which the left hand and upper lines are of medium weight, while the right hand and lower lines are drawn much heavier. In show drawings an elaborate border of original design is frequently used.

97. Visible Outline.

The visible outline shows those edges of the object which can be seen by the eye without interference by other parts of the object. The weight of the line, in general, is as shown. It must be varied, however, in some cases; for example, when two lines are close together it will be well to make the lines somewhat lighter in weight to prevent the ink running together.

98. Hidden Line.

The hidden line is used to represent such edges of an object as are not visible to the eye, but are hidden by some other part. It usually consists of dashes about 1/8 inch long with spaces a little less than half that length. In large drawings, where hidden lines will not interfere or obscure the clearness of a drawing, the hidden lines may be made of the same weight as the outline. In such cases it will be found an advantage, inasmuch as the same setting of the pen may be used for both outlines and hidden lines. The weight of such hidden lines and length of dashes will, however, depend upon the kind of drawing being made, the possibility of interference and the proximity of other lines.

99. Center Line.

The center line is used to designate the axis of a drawing. In many cases it may be used advantageously for the alignment of two views. In some drawings it is omitted altogether since the axis is obvious or self-apparent. It should be used in all drawings which are symmetrical about a center; such as cylindrical objects, holes, hubs, etc. In many drawings the center line is used as a guide from which to take important measurements. This is particularly true of machine drawings. The center line consists of a fine broken line of dashes about 1/2 inch long alternating with dashes about 1/16 inch long with a very short space between the dashes.

100. Projection Line.

The projection line, sometimes called the line of motion or construction line, is used to project or carry definite points from one view to another. It is almost always a horizontal, vertical, or circular line. It may be oblique, however, when a view or section is taken in such positions as to be inclined to the planes of projection. In practical drawing the projection line is very rarely used; when used, it is for construction purposes only and is never left on the finished drawing. The beginner should make use of projection lines until he thoroughly understands the science of projection. Projection lines consist of fine dashes about 1/16 inch long, with spaces of less length than the dash, when drawn in ink. When drawn in pencil the line is usually very fine and continuous.

101. Auxiliary Line.

The auxiliary line, also called assisting line or construction line, is used as a help in obtaining some point in construction. Illustrations of its use will be seen in the various figures. It is usually omitted from the completed drawing except in geometrical problems and in such drawings as are used for reference.

Auxiliary lines are also used to indicate the position of a cutting plane showing where a section is taken. It consists of a fine dash about 3/8 inch long, followed by two dashes, 1/16 inch long, with very short spaces between the dashes.

102. Extension Line.

The extension line is used as a projection for dimension lines, when the dimension is to be placed in a more convenient position outside of a view. It consists of fine dashes about 1/4 inch long, with very small spaces intervening.

103. Alternate Position Line.

The alternate position line is used to indicate a new or alternative position of some part of a machine, which has movement; as, for instance, the alternative position of a lever, or a piston in motion. It should consist of fine dashes about 3/16 or 1/4 inch long.

104. Cross-hatching Line.

Cross-hatching lines are used to show that a surface is cut and an end section is exposed to view. Such sections are "hatched" with very fine continuous lines, sometimes dotted lines, about 1/16 inch apart. The spacing will, however, depend somewhat upon the extent or size of the surface to be covered. The lines are generally drawn at an angle of 45°.

105. Dimension Line.

Dimension lines are used to indicate the direction in which a measurement is taken. The line always terminates with an arrowhead at each end, indicating the limit of the dimension. Sometimes the figures are placed on or above these lines. This latter practice, while it saves a little time, is oftentimes likely to lead to a misinterpretation and therefore should not be followed. The proper use of the line is shown in the examples.

When the distance to be dimensioned is short, the line on each side of the figures may be continuous. If the distance is long, it should be broken as illustrated.

106. Light-ray Line.

This line, showing the projection of a ray of light, is used in the casting of shadows of objects upon the planes of projection. Since a ray of light can not be shown as such, its representation upon the planes is indicated by projection, and such lines consist of very fine dashes about 1/2 inch long.

107. Break Lines.

Break lines are used to indicate that some part or portion of an object is broken away, for the purpose of showing its internal construction more clearly, than is possible with the aid of hidden lines.

The line illustrated on the left is generally used for metal breaks, while the more irregular line on the right is used for showing wood breaks. In the examples a number of illustrations of the uses of these lines are shown.

DIMENSIONING

108. When a drawing is made as a guide for the purpose of executing an object in wood, metal, or other material, it should not only be drawn to a suitable definite scale, but should also contain all the necessary dimensions and explanatory notes, which a workman may need for the purpose of making the model represented.

To dimension a drawing, means to place on it such measurements as a workman may need for the purpose of constructing the object represented by the drawing.

109. Drawings intended as working drawings should be either made full-size or to some reduced scale and properly dimensioned, since no good workman relies on the accuracy of the drawing to construct his object, but directly on the given dimensions. Dimensioning is especially necessary on inked drawings which can not be accurately scaled on account of the wide lines used. The difficulty of making accurate measurements is increased if shade lines are employed.

110. In placing dimensions on drawings, the aim should be to give only those which the workman must have, and no more. All objects are measured in three dimensions, and the dimensions should be placed in suitable places. It is not customary, in good practice, to put the same dimensions on more than one view of the same object. One or possibly two of these dimensions must appear on one view of the object, while the third dimension must necessarily be placed on one of the other two views, if three views are made. A repetition of dimensions is confusing and may lead to error.

111. The art of dimensioning and properly describing a drawing by explanatory notes, is very exacting and difficult, and requires a large amount of patience, practice, experience and general knowledge. There are no fixed rules which can be laid down, since no two drawings of the same object made by different men are dimensioned in exactly the same way, and yet both may be sufficiently correct.

It will be well for the beginner to study carefully the good examples of experienced draftsmen, and profit by their extended knowledge and practice.

112. Figs. 56 and 57 show methods employed by draftsmen for dimensioning horizontal, vertical and oblique distances; also

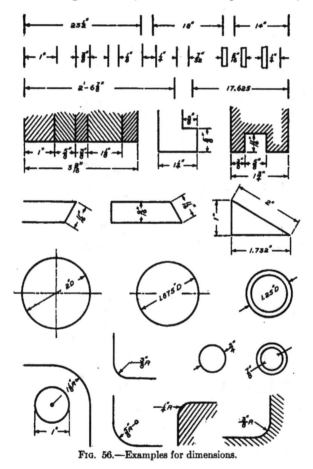

Fig. 56.—Examples for dimensions.

diameters, radii and fillets.   Practice varies considerably in the placing of dimensions.   Some draftsmen place all dimensions outside of the views, while others place them inside.   When a drawing is more or less complicated it is good practice to place some dimensions outside and others inside. Fig. 60 is an illustration showing a suitable distribution of dimensions.

In drawings for small objects when the dimensions are understood to be in inches, the sign '' may be omitted as in Fig. 60.

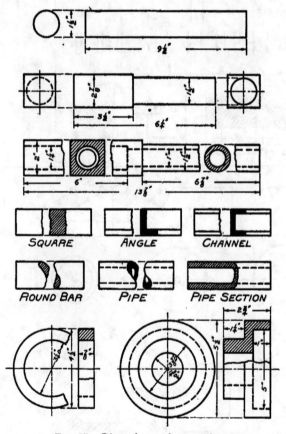

FIG. 57.—Dimensions and conventions.

When feet and inches occur they should be written 2′–6″ or 2 ft. 6″; the first method is the one most generally used. When the dimension is feet and no inches it should be written 3′–0″ or 3 ft. 0″.

113. Rules to be Remembered.

1. All dimensions indicating horizontal measurements should be parallel with the lower edge of the drawing, that they may be read from a position directly in front of the drawing.

2. Dimensions indicating vertical measurements should read from right to left.

3. If the upper end of an inclined line leans toward the right, the dimensions should read from the right. If it inclines toward the left, it should read from the left.

4. Dimensions up to, and including 2 feet should be given in inches; if over 2 feet, they should be indicated as feet and inches.

5. Arrow points should be placed at both ends of a dimension line.

6. Do not place more than one dimension on any one line.

7. Never place a dimension on a center line, or on a contour line of the drawing itself.

8. Do not place a dimension on the intersection of two lines.

9. If a dimension must be placed on a cross-hatched surface, leave an open space in the hatching for the figures.

10. Over-all dimensions should always be given, as well as detail dimensions of parts.

11. If an object is symmetrical the dimensions should be given from the center line.

12. Do not crowd a dimension into a space which is too narrow for the figures.

13. Always place a dash between feet and inches.

14. When a measurement is to be taken with great accuracy, the fraction of an inch should be given in decimals.

15. Over-all dimensions for patterns, furniture, and architectural work should always be given in feet, inches and fractions of an inch.

16. All figures should be neatly printed and not hurriedly written.

17. Do not place a dimension on a line which is inclined to both planes of projection.

18. Dimensions should, in general, be placed outside of the drawing, unless there is no possibility of a misinterpretation when they are placed inside.

19. When dimensions are placed outside of a drawing, extension lines should be drawn.

20. Use every precaution to secure accuracy in dimensioning every piece entering into the construction of the object represented.

21. The dimensions must indicate the actual size of the piece itself, regardless of the scale to which the drawing is made.

22. A good general size for figures employed to indicate dimensions is not over 1/8 inch high, and for the smaller work about two-thirds that size.

CROSS-HATCHING OR CROSS-SECTIONING

114. To cross-section a piece means to represent graphically the nature of the material from which the piece is to be made. This is accomplished by a variation in the character of lines

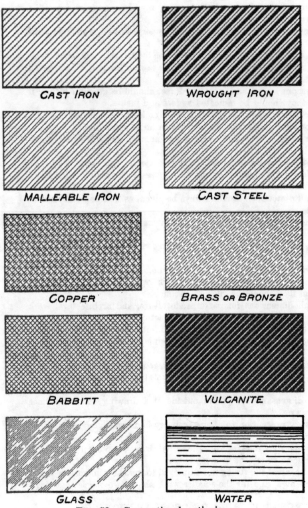

CAST IRON WROUGHT IRON

MALLEABLE IRON CAST STEEL

COPPER BRASS OR BRONZE

BABBITT VULCANITE

GLASS WATER

FIG. 58.—Conventional sectioning.

employed. The lines for cross-sectioning should be somewhat finer than those used to show the outline of the object to be sectioned. They are usually drawn at an angle of 45° and about 1/16 inch apart for average size drawings. When two or more adjoining surfaces are to be sectioned, the angularity of the lines should be changed so that a distinction between the various pieces can readily be seen. (See detail in Fig. 56.)

Section lines are not drawn in pencil, but are put in directly with ink. The spacing may be done with a section liner (Fig. 27) or by eye. The latter method is generally adopted by the draftsman, as it saves the time necessary to place and adjust the instrument. Spacing by eye is to be commended, as it not only saves time, but also gives considerable training for the eye and technique for the hand. The beginner in drawing, however, usually experiences considerable difficulty in obtaining uniformity in spacing.

115. The cross-sections shown in Fig. 58 are a part of a large number which have been submitted to the American Society of Mechanical Engineers for adoption as possible standards. For a practical application, see Fig. 60 (*D*).

SECTIONS

116. Purpose of Sections.

If a plane be passed through an object at any angle, and the part of the object in front of the plane be removed, the surface of the remainder of the object is said to be in section, or is a sectional view. The cutting planes used are generally taken to be either horizontal, vertical, or transverse.

If an object is hollow, it is frequently shown in section so as to show its internal construction with solid, instead of hidden lines, thereby making the drawing easier to read and reducing the possibility of error to a minimum. Where the internal construction of an object is simple, such as a pipe or a drilled hole, sections are not generally made; but where the internal construction is complex, such as a valve or steam engine cylinder, one or more sectional views should be made, depending upon the complexity of the interior of the object. (See Figs. 59 and 60. Sections are also shown in Fig. 57.)

117. Longitudinal Section.

If a cutting plane be passed through an object parallel to its greatest length, the section is said to be a longitudinal section.

The cutting plane is dependent upon the position of the object and may be either horizontal or vertical. [See Fig. 60 (*D*).]

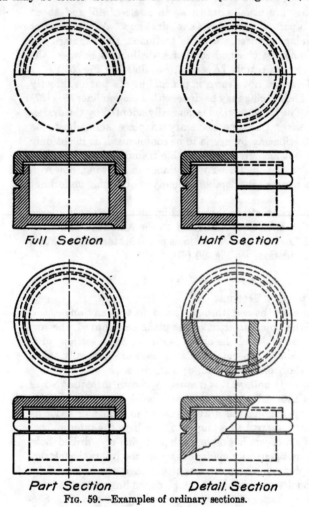

Fig. 59.—Examples of ordinary sections.

118. Transverse Section.

If the cutting plane be passed through an object in a direction at right angles to its length, the section is transverse. The cutting plane may be either horizontal, vertical, or oblique. [See Fig. 60 (*C*).]

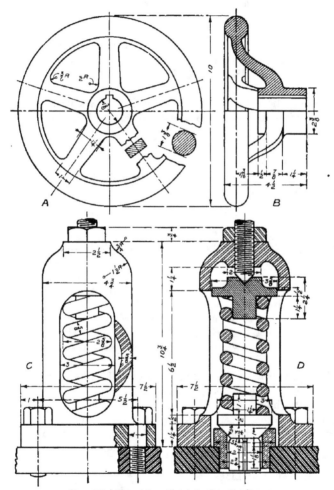

Fig. 60.—Examples of standard sectioning.

119. Oblique Section.

An oblique section is one which is neither longitudinal nor transverse.

120. Full Section.
If an object is cut completely in two, the section shown is a full or complete section. It may be horizontal, vertical or oblique. [See Fig. 59 also Fig. 60 (*D*).]

121. Half Section.
A half-sectional view is obtained by removing one-quarter of an object showing one-half in section and the other half full, with respect to a center line. [See Fig. 59 also Fig. 60 (*B*).]

122. Detail Section.
If any small portion of an object be removed to show a special detail of its interior, the section is called a detail section. [See Fig. 59, also Fig. 60 (*C*).]

123. Revolved Section.
A revolved section is a cross-sectional view of an object. It is usually placed directly on one of the regular views. [See Fig. 60 (*A*) and (*C*).]

SHADE LINES

124. Purpose of Shade Lines.
Shade lines are sometimes used on a mechanical drawing to give relief, thereby making it more attractive to the eye and in many cases easier to read.

There are no fixed rules laid down when shade lines should be used. Many draftsmen will apply shade lines to nearly all their drawings, while others never use them. Since it requires additional time to make a drawing with shade lines, their use is generally considered a waste of time, and they are therefore omitted.

Shade lines are always used in making patent office drawings, illustrations for catalogue work, and frequently on assembly, and on show drawings.

125. The conventional method in applying shade lines is to shade the lower and right-hand edges of all surfaces, assuming the source of light to come from the upper left-hand corner and at an angle of 45°.

126. Shading Straight Lines.
When straight lines are to be shaded, the shade is usually added to the outside of the line which bounds the surface. This, however, is not a fixed rule and the placing of a shade line will depend somewhat upon the drawing and upon the experience of the draftsman.

127. Examples of Shaded Straight-line Objects.

Fig. 61.—*a* shows one view of a rectangular object with a

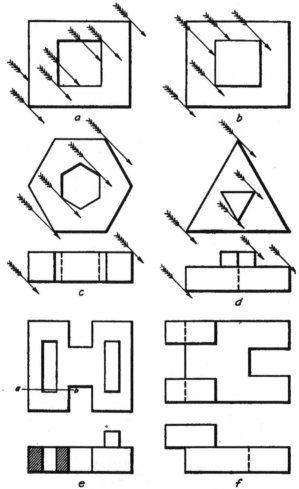

FIG. 61.—Examples of shaded straight lines.

square hole; the hole is indicated by shading the upper and left-hand lines. The surface of the object being in the light, the upper and left-hand surfaces of the hole are in the shade.

b shows a rectangular object with a square block placed on top. The lower and right-hand surfaces of the block being in the shade, the lines must be shaded as shown.

c illustrates a hexagonal figure with a hexagonal hole. The lines to be shaded can easily be determined by the arrows passing the corners as shown.

d shows a triangular object with a triangular projection. The proper edges to shade are determined on the drawing by the arrows.

e shows an object of irregular outline. The left-hand rectangle is a depression shown on the plan by the shade lines. The front view shows a section through the line *ab*.

f is a somewhat similar figure with two projections at the left. The proper location of the shade lines can be determined by drawing 45° lines passing the various corners, or simply by determining by the aid of the 45° triangle which surfaces are in the light, and which are in the shade.

128. Shading Circles and Circular Arcs.

In shading circles the outline is first drawn with the regular contour line. If the circle is to show the outline of a cylindrical surface, the point of the compass is moved down and toward the right, on a short 45° line, drawn through the center point to such distance as will give the thickness of the required shade. An arc is then drawn in shade line, and if the work is accurately performed, it will merge into the outline at the intersection of the circle and a 45° line, drawn up and to the right, passing through the center of the circle. The intervening space between the arc and circle, if the shade line is to be heavy, can be filled in by springing the compass or by slightly altering the radius.

If the circle to be shaded is to represent the internal cylindrical surface, the center of the compass should be moved up and to the left, on a 45° line, so that the shade will be shown on the outside of its circumference.

129. Shading a circle may also be accomplished by using the center point of the circle and gradually springing, or increasing the pressure on the compass until the heaviest part of the shade is reached, and then gradually decreasing the pressure to the end of the shade. This method requires considerable practice and skill and is not recommended for beginners. The method of using a new or second center is more certain to meet with success, and therefore is to be favored.

130. Examples of Shaded Circles and Circular Arcs.

Fig. 62.—*a* and *b* show how the shade lines are applied to circles denoting the interior and exterior of cylindrical surfaces.

c and *d* illustrate the shading of circular arcs and tangent

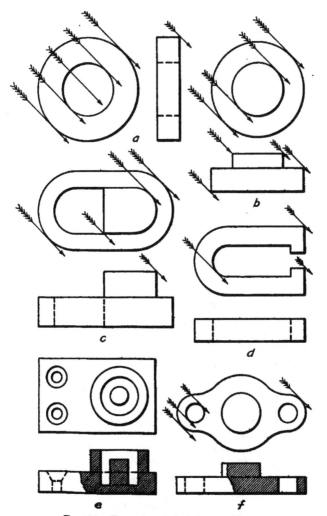

Fig. 62.—Examples of shaded curved lines.

lines. By reference to the front views it will be seen that hidden lines are not shaded.

e is shown in part section and illustrates the shading of lines which would be invisible if they were not shown in section. Since the bevel of the countersunk screw holes is an angle greater than 45° with the surface, as is shown in the front view, they are shaded in the plan.

f shows an object whose outline is composed entirely of circular arcs. The portion shaded is determined by the 45° lines placed at its extremities.

131. Rules to be Remembered.

1. The rays of light are assumed to come from the upper left-hand corner of the drawing.

2. All light rays are parallel lines and the projection of their direction is an angle of 45°.

3. All views of an object are to be considered as top views and equally exposed to the light throughout their entire surface.

4. A shade line should always be from two to three times heavier than the outlines or contour lines of an object.

5. Shade lines are always applied to those edges which bound light and dark surfaces.

LINE SHADING

132. Purpose of Line Shading.

Line shading is sometimes used on display drawings, show drawings or patent office drawings to give a more realistic or pictorial effect to the object or machine represented.

Line shading, as is the case with shade lines, is not used very extensively by the draftsman on account of the time it consumes for proper executions. The successful shading of single and double curved surfaces is an art not generally possessed by the average draftsman. It requires extensive practice and extreme care to obtain pleasing results.

133. Examples of Line Shading.

Fig. 63.—*a* and *d* show two surfaces which are parallel to the plane of the paper. They are shaded by lines drawn at equal distances apart and of uniform width. In *b* and *c* the surfaces shaded are inclined to the plane of the paper. To obtain this effect the spacing is uniform, but the width of the lines themselves is increased or diminished, as the figure may require. This is accomplished by uniformly opening or closing the pen for each line drawn.

d, e and f show surfaces shaded by vertical lines. The method for obtaining these results is the same as described in the foregoing examples.

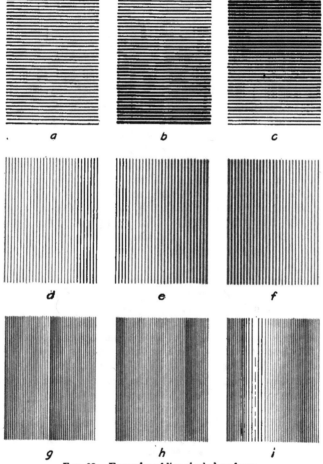

Fig. 63.—Examples of line-shaded surfaces.

g illustrates adjacent surfaces, as in a square prism. On the surface to the left the darkest part is farthest removed from the corner nearest to the eye; while the surface on the right is darkest nearest the eye.   This is the conventional method of shading inclined surfaces.

h shows three sides of a hexagonal prism.   The center surface of the object shows a uniform tone, as this surface is parallel to the plane of the paper.   The two outer surfaces are inclined to the paper and are shaded as previously described.

i is the shading adopted for a cylindrical surface.   **The light and dark parts** will be explained in the next example.

134. Examples of Shaded Surfaces.

Fig. 64.—*a* illustrates the plan and elevation of a half cylinder. The brilliant part of the shaded surface is found by drawing the line *ab* at an angle of 22–1/2° to the vertical center line, and projecting down upon the front view to the point *c*.' This will be the point where the angle of reflection is equal to the angle of incidence, consequently where the reflected rays return directly to the eye.   The darkest part is at that point where a 45° line, drawn from the center *b* of the plan, intersects the circumference of the circle at *d* and is projected down to *e* on the front view.

b shows the concave surface of a hollow cylinder.   This is usually shaded darkest at the edges.   The brilliant part is found by drawing a 22–1/2° line through the center, intersecting the surface of the plan at *a*, and projecting *a* down to *b*, on the front view.

c shows the front surface of the hexagonal pyramid shaded with a uniform tone, while the inclined surface lines are graded to produce the effect of light and dark as explained.

d shows a cone, whose brilliant and dark parts on the surface are found as in the cylinder.   In the figure all the shade lines meet at the apex of the cone, which convergence necessarily produces some very dark spots.   These may be avoided if the shading is produced by drawing the lines parallel to the edges of the cone, in which case, however, it will not give very good results.

e shows the shaded surface of a cylinder head.   The effect of a highly polished surface is produced as shown.   The flange which contains the bolt holes is not polished, therefore it is shaded by a number of concentric circles equally spaced.

f shows a sphere, which is a double curved surface.   The point of brilliancy may be accurately found by descriptive geometry, but it will be accurate enough for all practical purposes to assume it to be a distance from the center, equal to about one-third the radius, on a line drawn at an angle of 45° from the

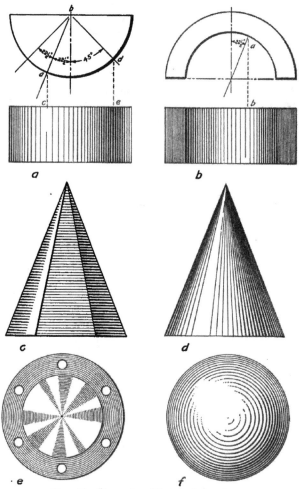

FIG. 64.—Examples of line-shaded surfaces.

center point upward, and toward the left, as in the cylinder. The darkest part is not at the edge, but should be drawn at a distance about one-fourth of the radius from the edge.   The effect is produced by using two centers for each arc drawn. This result may also be obtained by springing the compass as previously described.

In the examples shown, the surfaces are all shaded with unbroken lines, which however is not necessary, since excellent results may be obtained by a judicious shading in moderation. Several examples of this method are given in Fig. 153.

CHAPTER VI

PROJECTION

135. Introductory.

This chapter deals with the subject of projections as briefly as is consistent with clearness. A number of necessary definitions are given, a short description of perspective projection is included, but the emphasis is laid on orthographic projection. The first and third angle projections are discussed with emphasis on the latter, since this is most commonly used in commercial work.

The following are abbreviations used in projection drawing:

V designates Vertical Plane.
H designates Horizontal Plane.
P designates Profile Plane.
G L designates Ground Line.
V G L designates Vertical Ground Line.
Aux.G L designates Auxiliary Ground Line.

DEFINITIONS

136. That point in which a straight line passing from a point in space pierces a plane, is called the *projection of the point upon the plane,* called briefly *projection.* The line is called the *projection line,* and the plane is the *plane of projection.* Projection lines are called *projectors,* which may be drawn from various points on a line, plane, or solid to the plane of projection.

137. Projectors may be drawn at any angle with the plane of projection. When the projectors diverge from a point located in front of the plane of projection, they resemble a *cone of visual rays.* This form of projection is known as *Radial Projection,* or *Scenographic Projection,* or *Perspective Drawing,* and the plane upon which projections are made is called the *Vertical Plane* or *Picture Plane.*

138. If the projectors are parallel to each other and inclined to the plane, of projection, the form is called *Oblique Projection.* Under this classification, if the plane of projection is vertical, we have such types of projection known as *Cavalier Projection,* and *Cabinet Projection.* Drawings made in oblique projection will

71

give a pictorial effect closely resembling perspective. They are used for illustrating machine details and architectural construction, in preference to true perspective, on account of the ease with which solid forms and details can be drawn.

139. If the plane is horizontal and the projectors inclined, we have what is known as *Military Projection*, sometimes called *Military Perspective.*

140. If the projectors are parallel to each other and perpendicular to a horizontal plane, the form is called *Horizontal Projection*. The principal application for this type of projection is in topographical drawing.

141. Projection of a Point.

That point in which a straight line passing from a point in space to a plane, pierces that plane, is called the projection of the point upon the plane.

142. Projection of a Straight Line.

If the extremities of a given straight line in space are projected upon a plane and these projections connected by a straight line, the line so produced is the projection of the given line upon the plane.

143. Projection of a Curved Line.

If the projections of a convenient number of points, from a given curved line in space, upon a plane, are joined by a smooth curve, the curve is called the projection of the curved line.

144. Projection of a Plane Figure.

If a number of points from the outline of a plane figure are projected upon a plane and connected by lines, the figure thus obtained is the projection of the figure in space, upon the plane.

If the figure is bounded by straight lines, it will be necessary to project only the vertices or connecting points of the lines.

145. Projection of a Solid.

The projection of a solid upon a plane may be found by projecting a number of points from the surface of the solid to the plane and connecting such projections or points. If the solid consists of a number of plane faces it will be necessary to project only the corners or meeting points of the various lines.

PERSPECTIVE PROJECTION
(Fig. 65)

146. This illustration represents a simple object in space with a number of *visual rays* converging or meeting at a point. A plane is placed between the object and the point where the rays meet, and a picture, produced by the piercing points of the visual rays, is seen on the plane. The size of this picture is determined by the distance at which the plane is placed from the object. It would diminish in size if the plane were moved toward the point of convergence or meeting-point of the rays; and, on the contrary, it would increase in size if the plane were moved toward the object, the position of the point and object remaining unaltered.

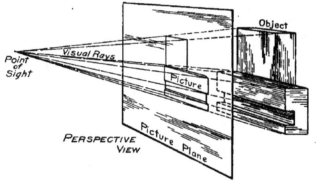

Fig. 65.—Perspective projection—showing the point of sight at a finite distance from the plane.

147. Since the rays are divergent from their meeting-point, they are not perpendicular but are inclined to both the object and the plane. The angularity of the rays depends upon the location of the point with reference to the plane. The location of the point may be so chosen that more than the front face of the object is seen and projected, or reflected, upon the plane. Pictures drawn by this method are known as having been drawn in perspective.

148. If the object and plane both remain in their fixed positions, the size of the picture on the plane may be increased by removing the point of convergence farther away, in which case the rays will change in angularity with the plane. If the point of convergence is removed to an infinite distance from the plane the rays are assumed to be no longer divergent from the point, but parallel. The point of convergence of a cone of visual rays is called the *point of sight;* and a line drawn from the point of sight perpendicular to the plane is called the *central visual ray,* while the piercing point of the central visual ray with the plane, is the *center of vision*.

149. In making working drawings, the point of sight is considered to be so far removed from the plane of projection that all rays are said to be parallel, and their direction perpendicular to both the plane and the object.

ORTHOGRAPHIC PROJECTION

150. Orthographic projection is also called *Perpendicular* or *Rectangular Projection*. The word "orthographic" is derived from two Greek words, *orthos* meaning straight and *grapho* meaning write. The compound word applied to drawing means straight-line drawing.

151. *Orthographic projection is the method of representing an object by its images upon two or more planes, accomplished by parallel lines of projection, drawn from various points of the object and perpendicular to the planes.*

152. For the complete representation of an object by orthographic projection, at least two views of that object are necessary. If the object is complex, three and sometimes more views will be required to show its entire form and details.

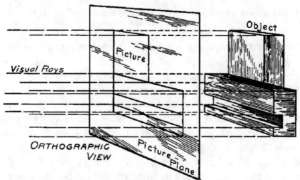

Fig. 66.—Orthographic projection—the point of sight being at an infinite distance from the plane.

153. Orthographic Projection of an Object.

Fig. 66 is another illustration of the object shown in Fig. 65 with a number of parallel visual rays perpendicular to the surface of the object. The plane, which is parallel with the face of the object, is cut or pierced by the visual rays passing through to the object, thereby producing or reflecting an image or picture of the object upon the plane. This picture is called the *orthographic projection of the object.*

Since the rays are parallel and perpendicular to both the object and the plane, the size of the projection upon the plane accords perfectly with the proportions of the object, and the distance between the plane and object will not, in any way, change the size of the projection.

It will be observed that the thickness or depth of the object can not be shown upon the vertical plane. If it is desired that the thickness should be shown, a horizontal or end plane must be employed, through which visual rays must be passed; the resulting view will be an image of a picture in which the thickness is visible.

If the visual rays in any projection are parallel and perpendicular to the object, and if the picture plane is parallel to the face of the object, the visual rays will be perpendicular to the plane, and the resulting projection upon the plane will be an exact reproduction of the view of the object in its exact size and proportions.

154. First Angle Projection.

Two planes which intersect each other at right angles form four dihedral (two-sided) angles. Mechanical drawings may be projected in any one of these four angles; but it is a convention agreed on by draftsmen to use either the first or third angle. In descriptive geometry all four angles may be used.

155. An illustration of first angle projection is shown in Fig. 67. Note first the point A in space, and the projections of this point on three planes. The plane marked H is the horizontal plane, and V is the vertical plane. The line of intersection of these two planes is the ground line. The plane marked P is the profile plane and is perpendicular to both the horizontal and vertical planes. The lines where the profile plane intersects the horizontal and vertical planes, are called axes.

The representation of point A in space has been located upon the planes by drawing perpendicular lines from the point to the planes (projections of the point). The three projections are designated as a_h, a_v and a_p, the letters h, v, and p signifying horizontal, vertical and profile projection.

As drawings are made on a flat surface (or plane), it follows that it is not practical to represent a point on the three planes in their normal position; therefore these planes must be revolved into one plane. The three projections are then shown on one surface. To make this possible the profile plane is revolved about its vertical axis until it coincides with the vertical plane, by curving the projection of the point A, from a_p, to a'_p. The vertical plane is then revolved about the ground line until it coincides with the horizontal plane. The vertical projection a_v will revolve with the plane and fall upon a'_v, and a'_p will fall upon a''_p.

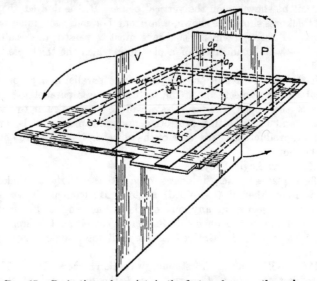

FIG. 67.—Projections of a point, in the first angle, upon three planes.

156. By referring to Fig. 68, which shows the planes in their revolved position, the projections of the point A will be found designated as a_h, a_v and a_p.

It must be clearly understood that these planes have no appreciable thickness and that in reality such planes do not exist, but are imaginary. Therefore, in practice, one plane does not fall upon or superimpose another.

157. Fig. 69 shows the drawing board, paper, T-square and triangle in position, also the three projections of the point A in a true relation to the planes of projection.

158. The ground line may be drawn anywhere across the

paper, and the surface above it is the vertical plane, while the surface below is the horizontal plane. The surface to the right of a vertical line drawn from the ground line may be considered as being the profile plane.

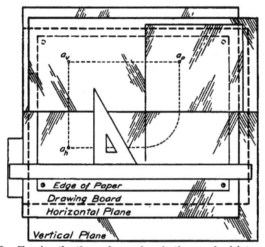

FIG. 68.—Showing the three planes of projection revolved into one plane.

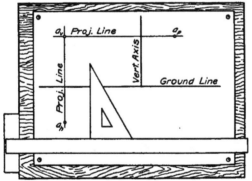

FIG. 69.—Showing the three projections of a point as the draftsman would see it.

159. By referring to the figures it will be seen that the distance of the vertical projection of the point above the ground line is identical with the distance of the point from the horizontal plane, while the distance below the ground line is the corresponding distance of the point from the vertical plane.

160. Third Angle Projection.

Fig. 70 shows a point A in space in the third angle. Its projection upon the horizontal plane is a_h, upon the vertical plane a_v and upon the profile or end plane a_p.

When the end plane is revolved about its vertical axis until it coincides with the vertical plane, a_p will fall upon a'_p, and if the vertical plane is then revolved about the ground line until it coincides with the horizontal plane, a_v will fall upon a'_v and

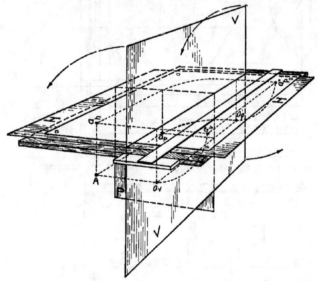

FIG. 70.—Projections of a point, in the third angle, upon three planes.

a'_p will fall upon a''_p. The revolved positions of the planes are shown in Fig. 71 and the projections of the point are designated by a_v, a_h and a_p.

161. By a careful study of Fig. 70 it will be seen that, in third angle projection, the vertical projection of the point necessarily falls below, while the horizontal projection is above the ground line. It will also show that the end plane may be revolved about either its horizontal or its vertical axis, and thereby be made to coincide with either the horizontal or the vertical plane. If the end plane is revolved about its vertical axis, the end projection of the point will be shown below the ground line, while in first angle projection it will be above.

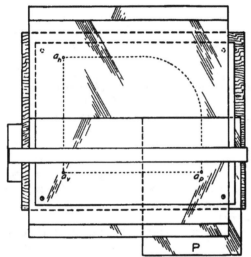

FIG. 71.—Showing the three planes of projection revolved into one plane.

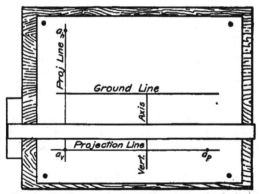

FIG. 72.—Showing the three projections of a point as the draftsman would see it.

The profile or end plane, in third angle projection, is generally revolved about its vertical axis, although occasions may arise when it may be desirable to revolve it about the horizontal axis, thereby making it coincide with the horizontal plane. This will show the end view of the point above the ground line, and in first angle projection it will be below.

162. Fig. 72 shows the drawing board, paper and the T-square in proper position, and as the draftsman would see it. Since the point A in this case is shown in third angle projection and the end plane is revolved about its vertical axis, the end projection of the point will appear below GL, as shown in the figure.

The three following explanations of the projections of a *point*, which are also true of a *line* or a *solid*, are of great importance to those who wish to master the science of mechanical drawing, and should therefore be carefully studied and compared with the illustrations.

163. *First.*—a_h being the projection of a point, line, or object in space, upon the horizontal plane, shows a view looking down upon the point, line, or object, in a direction perpendicular to the plane. This projection, in practice, is called *the plan view*, or simply *the plan*, in architectural drawing; in machine drawing, it is called *the top view*.

164. *Second.*—a_v being the projection upon the vertical plane, shows a view looking in a horizontal direction. The plane is supposed to be vertical and the horizontal visual rays, as explained before, are perpendicular to the plane. This view gives *the front elevation*, as it is called in architectural drawing; whereas in machine drawing, it is called *the front view*.

165. *Third.*—a_p being the projection upon the profile or end plane, shows a view looking in a horizontal direction and perpendicular to the plane. This view in architectural drawing is called *the end elevation*, while in machine drawing it is known as *the end view*.

In architectural drawing a view from each end of a building is generally necessary, thereby giving two end elevations. In machine drawing, unless the object is somewhat complex, one end view will generally give all the information required by a workman as a guide for constructing the object desired.

166. Projection upon Three Planes.

Fig. 73 shows a simple object placed on a drawing board with a vertical plane in position, somewhat removed from the object. The projection lines drawn perpendicular to the vertical plane, and connected by straight lines, show the outline or front view of the object, which in the illustration is parallel to the plane.

Fig. 74 shows the plan view drawn on the horizontal plane with the aid of vertical projection lines.

Fig. 75 shows the end view of the object produced in the same way as the front view.

167. Fig. 76 shows the three planes in position, as the draftsman would imagine them to be. The ground line AB is pro-

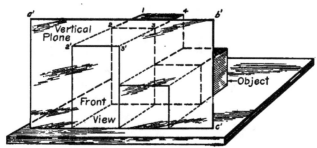

FIG. 73.—Showing the front view in orthographic projection.

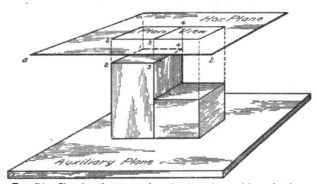

FIG. 74.—Showing the top or plan view in orthographic projection.

duced by joining the plane edges ab and $a'b'$ of Figs. 73 and 74. The vertical axis BC is obtained by joining the plane edges $b'c'$ and $b''c''$ of Figs. 73 and 75.

168. Referring to the figures, it will be seen that the projection lines passing from the object to the planes of projection are perpendicular to those planes; consequently, their representation upon the planes in Fig. 76 must necessarily be perpendicular to the lines AB and BC. This shows that the plan view must be placed directly above the front view and not moved either to the right or left from the position shown in the figure. It likewise shows that the end view should be moved out in a horizontal direction, and not up nor down. Either the plan or end views, or both, may, however, be moved closer to,

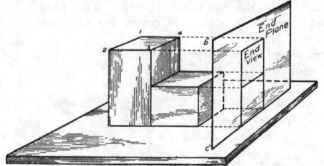

Fig. 75.—Showing the end view in orthographic projection.

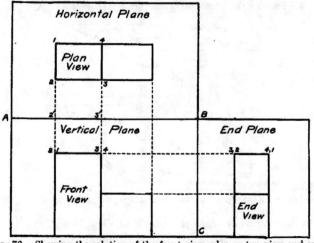

Fig. 76.—Showing the relation of the front view, plan or top view and end view when drawn on a flat surface.

or farther from, the front view without violating the rules of orthographic projection.

169. Points to be Remembered.

The lines drawn from an object to the planes of projection are called projectors. In orthographic projection such lines are perpendicular to the axis about which the planes are revolved.

The planes upon which the images or views of an object are drawn are called the planes of projection.

Generally three planes of projection are used:

(a) The horizontal plane, designated as H.

(b) The vertical plane, designated as V.

(c) The end or profile plane, designated as P.

These planes of projection are mutually at right angles, or perpendicular to one another.

The intersection of the horizontal with the vertical plane is called the ground line, designated as GL.

The intersection of the vertical plane with the end plane is called the vertical axis, designated as VGL.

Since the planes of projection must be represented on the drawing paper, they are necessarily on the same flat surface.

The horizontal and vertical projectors of the same point meet at the ground line, and are therefore in the same straight line.

The image or view drawn upon the horizontal plane, is the plan or top view.

The image or view drawn upon the vertical plane, is the elevation or front view.

The image or view drawn upon the end plane, is the end elevation or end view.

In third angle projection, the horizontal plane is above the ground line and the vertical plane is below.

The end plane in third angle projection, is usually assumed at the right of the vertical plane, resulting in the vertical axis being below and perpendicular to the ground line. In some cases the end plane may be placed above the ground line, in which instance it is at the end of the horizontal plane.

170. Projections of Lines.

The true length of a line in space, is shown by its projection upon that plane to which it is parallel.

171. Fig. 77 shows a line AB in the third angle with its end B touching the vertical plane. The view, on the vertical plane, appears as a point, since the line is perpendicular to that plane. The view upon the horizontal plane is a line, and since the line, in space, is parallel to the plane, its horizontal view is equal in length to the line itself, thereby showing the true length of the line in its horizontal projection.

The line CD in the same figure is removed from both the horizontal and vertical planes, but its true length is shown in its horizontal projection, because, like line AB, it is parallel to the horizontal plane.

172. In Fig. 78 there are three lines all of which are parallel to the vertical plane, and their true lengths may be measured there by their vertical projections. Line EF touches the horizontal plane at E. Line GH is removed from both planes. Line IJ

rests upon a plane parallel to the horizontal plane, which is called an *auxiliary horizontal plane*.

173. Fig. 79(*a*), shows the end view of Figs. 77 and 78 placed end to end, and Fig. 79(*b*), shows the view of the horizontal plane and the revolved position of the vertical plane with the projections of the lines in space.

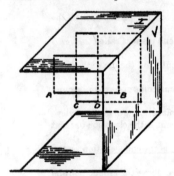

Fig. 77.—Orthographic projections of horizontal lines.

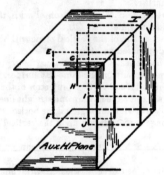

Fig. 78.—Orthographic projections of vertical lines.

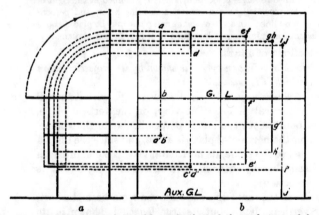

Fig. 79.—Showing the orthographic projections of the end, top and front views of horizontal and vertical lines.

174. A little study of the illustrations will show that since lines *AB* and *CD* are parallel to H and perpendicular to V, their H projections are perpendicular to the ground line, while their V projections are points. In the other case the lines are parallel to

V and perpendicular to H; therefore their V projections are perpendicular to the ground line, while their H projections are points.

175. Lines Parallel to One Plane.

If a line is parallel to one plane and inclined to another, its true length will be shown upon the plane to which it is parallel.

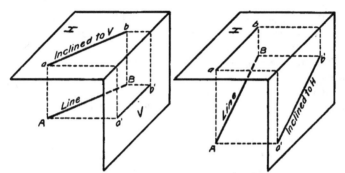

FIG. 80.—Illustrating a line parallel to H, and its projections upon H and V.   FIG. 81.—Illustrating a line parallel to V, and its projections upon H and V.

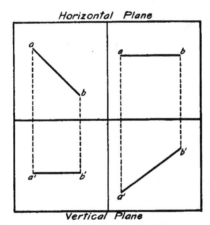

FIG. 82.—Showing the projections of inclined lines on a flat surface.

176. In Fig. 80 the line *AB* is inclined to V, and parallel to H, *a′b′* being its vertical projection, and *ab* its horizontal projection. Lines drawn from *A* and *B* perpendicular to the plane will form a polygon *Aa′b′B*, in which the angles at *a′* and *b′* are right angles.

The projector Aa' being greater than Bb', the polygon is not rectangular; therefore the projection as represented by $a'b'$ is foreshortened. Since the line is parallel to H, $AabB$ will form a polygon in which the angles at a and b are right angles, and the projectors Aa and bB being of equal length, the projection ab will therefore be equal to the line AB.

177. In Fig. 81 the line AB is parallel to V and inclined to H. Its true length will therefore be shown upon V, while its projection upon H will be foreshortened.

Fig. 82 shows the projections of both lines as they will appear when drawn upon the paper.

178. Lines Inclined to Two Planes.

The projections of a line which is inclined to two planes will be foreshortened upon both planes. The line must, therefore, be

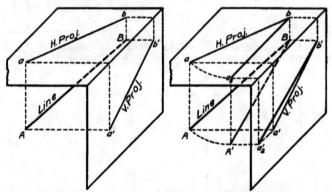

Fig. 83.—Illustrating a line inclined to H and V, and its projections.

Fig. 84.—Illustrating a line inclined to H and V, its projections and revolved position.

revolved in such a manner as to place it parallel to one of the projecting planes, if its true length is to be measured.

179. In Fig. 83 a line AB is inclined to both H and V and its projections are ab and $a'b'$. If the line is turned about B as a center until it becomes parallel to V, and the end A is neither raised nor lowered, A will move to A' (Fig. 84), and its H projection will then become ba_2, while its V projection will become $b'a'_2$, the point a' having moved to a'_2 in a direction parallel to the ground line. Since the line AB is now parallel to V, its true length can be measured upon that plane, and is shown by $b'a'_2$.

Fig. 85(a) shows the line *AB*, also its revolved position when placed parallel to V, as the draftsman would see it. Fig. 85(b) shows the same line, and also its revolved position when placed parallel to H. Either of the two methods will give the required length.

Sometimes it is more convenient to revolve a line by placing it parallel to V, instead of H, and sometimes the reverse is more

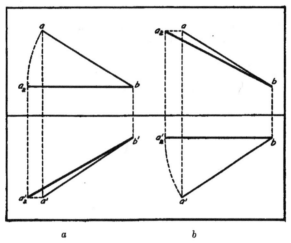

a	b

Fig. 85.—Showing the projections of inclined lines. (a) Inclined line revolved to a position parallel to V; (b) inclined line revolved to a position parallel to H.

convenient; which of the two methods would be most suitable to use is dependent entirely upon the problem to be solved.

Note.—When the H projection of a line is parallel to the ground line, the V projection of the line shows its true length.

When the V projection of a line is parallel to the ground line, the H projection of the line shows its true length.

180. Rules to be Remembered in Orthographic Projection.

All problems in this book are solved in third angle projection.

1. Any projection of a point is a point.
2. The projection of a straight line may be a line or a point.
3. A vertical line appears in its true length on the vertical plane.
4. A horizontal line appears in its true length on the horizontal plane.
5. When a line is parallel to one plane of projection it appears in its true length on that plane.

6. When a line is inclined to a plane of projection, its projection on that plane will be foreshortened.

7. The projections of a line oblique to both the horizontal and vertical planes will be foreshortened, and, therefore, will not appear in its true length on either plane.

8. If the horizontal and vertical projections of an inclined line are perpendicular to the ground line, its true length appears upon the end plane.

9. A line inclined to the three planes of projection must be revolved and placed parallel to one plane, before its true length can be found.

10. The projection of a plane figure upon any plane may be a line, or a figure.

11. The projection of a solid upon any plane is a figure.

12. In third angle projection, the distance of a point above the ground line measures its distance from the vertical plane, while the distance of a point below the ground line measures its distance from the horizontal plane.

13. The position of an object, relative to the planes of projection may be changed, while its projections remain unaltered.

CHAPTER VII

MECHANICAL DRAWING PRACTICE

SECTION I

INSTRUMENT EXERCISES

181. Plates Nos. 1, 2, 3 and 4 present studies for preliminary practice, to enable the student to acquire the use of the T-square, triangle, scale, dividers, pen, and facility in the use of the compass. Skill in the manipulation of the various instruments is a first requisite of the draftsman. The beginner should strive, at the start, to draw *clean-cut*, uniform lines, both straight and curved, and to make lines of exact and definite length.

In these drawing exercises, preparatory to inking, no more pencil lines should be drawn than are necessary. The preliminary pencil lines must be made quite fine, but dark enough to be seen clearly without straining the eyes. When the exercise is completed in pencil, the superfluous lines should be erased and those that are to remain should be retraced with a somewhat softer pencil. Every detail in each exercise must be clearly shown in pencil before beginning the inking-in process; this avoids the necessity of making erasures and corrections, which are difficult to make when the exercise is finished in ink.

To lay out the sheets for drawing the line exercises, proceed as follows:

1. Draw the border lines as explained in Chapter III.
2. On left vertical border line lay off the following distances, beginning at its lower end, 5/8", 3-1/4", 1/2", 3-1/4".
3. With the T-square, draw horizontal lines through the points.
4. On lower border line lay off distances, beginning at the left, 7/8", 3", 5/8", 3", 5/8", 3".
5. With the triangle placed against T-square, draw vertical lines through the points. This will give six rectangles forming the outlines of the exercises.
6. Erase the lines connecting, also those extending beyond the rectangles.
7. Re-measure the spacings to make sure they were correctly made.
8. With a somewhat softer pencil, draw the outlines so that they may be clearly seen.

Before beginning any exercise the student should carefully study the layout drawing shown above the completed example.

PLATE 1

LINE EXERCISES

182. Fig. A.—This exercise offers practice in drawing horizontal lines of equal length and uniform spacing, by the aid of the T-square.

Divide the height into twelve equal parts and through the points of division draw the required lines.

NOTE.—To divide a line or space into equal parts, see GEOMETRICAL PROBLEMS, Figs. 94 and 95.

Fig. B.—This exercise offers practice in drawing vertical lines of equal length and uniform spacing, by means of the T-square and triangle.

Divide the length into ten equal parts and through the points of division draw the required lines.

Fig. C.—This exercise requires the joining of a number of vertical, to an equal number of horizontal lines.

First draw a diagonal line from the lower left-hand to the upper right-hand corner of the rectangle and divide it into twelve equal parts. From the points of division draw the horizontal lines to the diagonal, then draw the vertical lines.

Fig. D.—This exercise requires the drawing of uniformly spaced lines inclined toward the right at an angle of 60°.

Divide the diagonal of the rectangle into fourteen equal parts. Draw the required lines with the 60° triangle resting against the T-square blade.

NOTE.—These lines should be drawn upward.

Fig. E.—To draw uniformly spaced lines inclined toward the left making an angle of 60° with the horizontal.

Divide the diagonal of the rectangle into fourteen equal parts. With 60° triangle draw the required lines.

NOTE.—These lines should be drawn downward.

Fig. F.—Exercise in drawing lines inclined at angles of 60° in both directions, and meeting on a pencil line drawn horizontally through the center of the figure.

Draw a fine horizontal line through the center of the figure; then draw diagonals. Divide each diagonal of the two rectangles into twelve equal parts. Draw the required lines with the 60° triangle. *Time, 3 hours.*

PLATE 1.

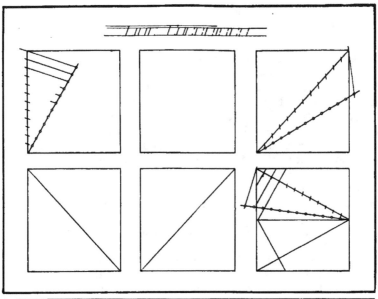

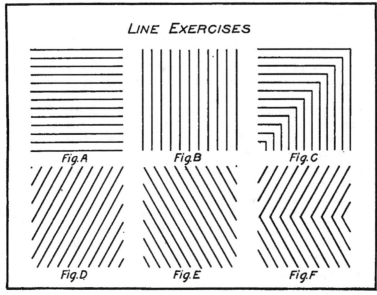

PLATE 2

LINE EXERCISES

183. Fig. A.—To draw a number of equally spaced horizontal lines, each consisting of four dashes.

Bisect the horizontal length of the figure and lay off on each side of the bisector a distance equal to 1/32 inch; through these points draw vertical lines. Bisect the distance between these vertical lines and the vertical sides of the rectangle, and lay off the same distances as described above, on each side of these bisectors, and erect vertical lines. Divide the vertical distance into twelve equal parts, and through the points of division draw the horizontal lines as shown.

NOTE.—This method of bisection will give dashes of equal length and spacing.

Fig. B.—Exercise in drawing vertical lines consisting of long and short dashes.

Lay off the following dimensions on the left vertical edge of the rectangle, beginning at the bottom: 7/8, 1/16, 3/16, 1/16, 7/8, 1/16, 3/16, and 1/16 inch. Through these points draw horizontal lines. Divide the horizontal distance into ten equal parts and draw vertical lines as shown.

Fig. C.—To draw horizontal and vertical lines of definite lengths and equal spacings.

Draw the diagonals, and with their point of intersection as a center, draw a circle of 1/8 inch radius. Tangent to this circle draw auxiliary lines on both sides and parallel to the diagonals. These auxiliary lines will give the limitations of horizontal and vertical lines required. Divide both the vertical and horizontal lengths of the figure into twelve equal parts, and through these points draw the required lines.

Fig. D.—To draw short horizontal and vertical lines of definite length, like a pattern of parquetry floor.

Divide one vertical and one horizontal side of the rectangle, each into twelve equal parts. Draw the three full-length horizontal and vertical lines as shown. The short lines may then be drawn their proper length.

Fig. E.—Exercise for drawing short diagonal lines of definite length, like a pattern for weaving.

Draw the two diagonals and divide each into twenty equal parts. Then with the triangle resting against the T-square blade, placed at an angle so that the inclined edge of the triangle is parallel to a diagonal, draw parallel lines through the points of division. Reverse the triangle and adjust the T-square and draw lines parallel to the second diagonal. The limitations of this second group of lines are determined by the first group of diagonals drawn, which indicate their proper length.

PLATE 2.

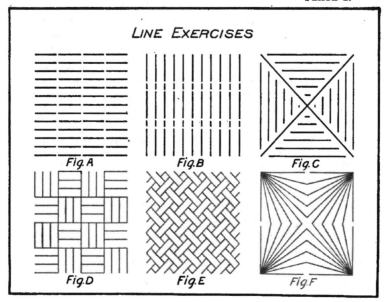

LINE EXERCISES

Fig. A Fig. B Fig. C

Fig. D Fig. E Fig. F

Fig. F.—To draw lines of varying angles, meeting on pencil lines drawn through the center of the figure, and terminating at the corners.

First draw horizontal and vertical bisecting lines through the rectangle. Divide each bisecting line into nine equal parts. From these points of division draw lines to the corners.

Time, 4 hours.

NOTE.—When inking-in, draw *from* the corners instead of *toward* them, to avoid blotting.

PLATE 3

SPACING AND LINE EXERCISES

184. The figures on this plate are intended as exercises preliminary to line shading of single-curved surfaces, and as a training for the eye in spacing, by gradually increasing and decreasing distances. The beginner will find that considerable skill and practice are required to produce a surface with a gradually changing tone value. The exercises are first to be drawn in pencil and all necessary corrections made before the inking-in is begun.

Fig. A.—Draw an arc of 90°, having its center at either extremity of the lower line and of a radius equal to the height of the surface to be spaced. Divide this arc into as many equal parts as there are to be lines in the surface of the figure. Draw horizontal lines through the points found, thus producing the lines required. Extreme accuracy in workmanship will be required to produce good results.

NOTE.—Arcs of circles may be subdivided by trial divisions with the aid of the dividers, or they may be divided with the aid of the protractor, by dividing the number of divisions required into the number of degrees in the given arc, which will give the number of degrees in each part.

Fig. B.—In this figure the center of the dividing arc must be taken at one extremity of the upper line. This arc is then divided and lines drawn as in the foregoing figure.

Fig. C.—This figure requires a half circle with its center at the middle point on either of the vertical bounding lines, and of a diameter equal to the length of the line. The half-circle should be divided into the number of spaces required and lines drawn through the points found.

Fig. D.—The surface of this figure may be divided by drawing and dividing a half circle, having its center on a diagonal of the figure, the diameter of the circle being equal to the length of this diagonal. Divide this figure into twenty-four parts.

Fig. E.—This figure should be divided into twenty-four parts and the lines drawn as shown. The placing of the dividing half-circle is obvious.

Fig. F.—This figure should be divided by using a half circle as explained for the fourth figure. Divide the figure into forty parts.   *Time, 3–1/2 hours.*

PLATE 3.

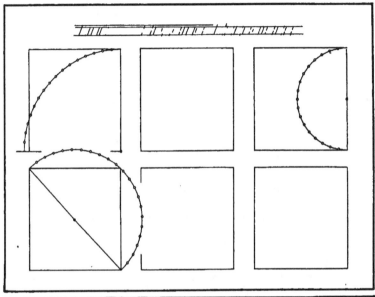

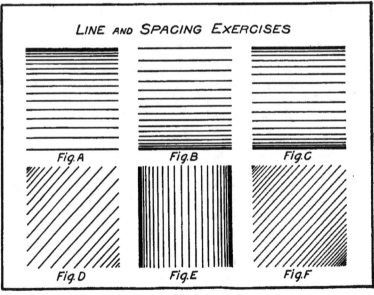

LINE AND SPACING EXERCISES

Fig. A   Fig. B   Fig. C

Fig. D   Fig. E   Fig. F

PLATE 4

CIRCULAR ARC EXERCISES

185. The first figure on this plate gives practice in the use of the compass in drawing a number of concentric circles and short circular arcs.

Bisect a horizontal and a vertical side of the rectangle and through these points draw center lines. With the intersection of these lines as a center, draw a circle 5/16 inch in diameter. On either center line prolonged, lay off with the rule, from the circumference of the 5/16 inch circle, six 5/16 inch spaces and through these points of division draw the required circles and arcs.

The second figure gives practice in drawing circular arcs of definite lengths.

Draw a diagonal through the rectangle, and with the rule lay off, beginning at either end, thirteen 5/16 inch spaces. Through the points of division draw the arcs.

The third figure gives practice in drawing arcs and short dashes.

Draw two center lines as explained for the first figure. On both sides of these center lines, lay off with the rule, two 1/16 inch spaces and draw lines parallel to the center lines. With the compass draw arcs and short dashes as illustrated.

The fourth figure gives practice in joining straight lines to arcs of circles.

From the center of one vertical side of the rectangle draw two lines making angles of 30° with the horizontal. On one line lay off eleven 5/16 inch spaces. With the compass draw arcs terminating on the oblique lines and with the 60° triangle draw straight lines from the extremities of the arcs.

NOTE.—When inking, first draw the arcs, then the straight lines.

The fifth figure gives practice in joining arcs to arcs, also to straight lines.

Divide a vertical center line into eleven equal parts. With the middle of the fourth space from each end, as a center, draw half circles, then draw the horizontal lines.

The sixth figure gives practice in joining semicircular arcs by straight lines.

Draw a vertical center line, also two horizontal lines 3/4 inch apart. Draw the semicircles as shown, and join by straight lines.   *Time. 4 hours.*

PLATE 4.

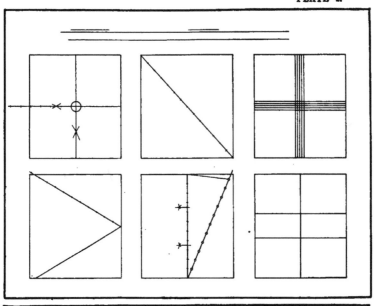

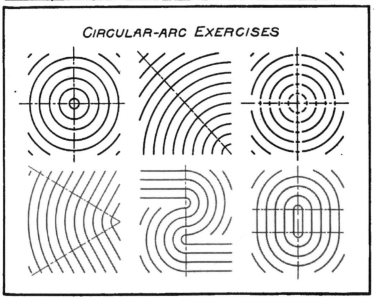

CIRCULAR-ARC EXERCISES

SECTION II

PROJECTIONS OF LINES, PLANE FIGURES AND SOLIDS

186. General drawing practice presupposes a knowledge of *Planes of Projection,* which is fundamental to all successful drawing work. · Drawing from models, or from drawings, or from instruction given by the teacher, will not give a student the ability to visualize, to conceive objects in space, or to think in three dimensions. This ability comes only from a mastery of the principles of projection. Possessing a knowledge of projection, the draftsman can make drawings without having recourse to a model, object, or drawing.

Elementary projection is not difficult to master, and the following nine problems offer the necessary basis of the theory, and sufficient drill for practice.

Before beginning the exercises the student should carefully study the chapter on Projection.

PROJECTIONS OF STRAIGHT LINES

187. Principles.

1. In the projection of a straight line, the drawing of that line may be either a straight line or a point, depending on the relation of the line to the plane of projection.

2. The projection of a straight line parallel to a plane of projection will have its true length shown upon that plane.

3. The projection of a straight line inclined to a plane of projection will appear foreshortened on that plane.

4. The projections of a line inclined to two planes of projection will appear foreshortened on both planes.

Illustrations of lines inclined to both horizontal and vertical planes are offered by the hip rafters of roof construction, or the corner or arris of a hopper-box.

NOTE.—The line formed by the intersection of two surfaces not in the same plane, as in prisms and pyramids, is technically called an *arris,* but commonly called a *corner* or an *edge.*

Before beginning any exercise the student should carefully study the layout shown below the completed example.

PLATE 5

VERTICAL LINE
(First Position)

188. Example.

A line AB in space is 3-1/2 inches long. It rests on an auxiliary horizontal plane which is 4 inches from H. In its first position on the Plate it is perpendicular to H and 3/8 inch from V. Since the line is perpendicular to H, its projection upon the H plane is a point ab, and since it is parallel to V, its true length appears upon the V plane and is shown as $a'b'$.

The second position on the Plate shows the projections of the line when inclined to H and parallel to V. The line has been revolved parallel to V about its end A, making an angle of 45° with H. Since the line remains parallel to V, its V projection is not altered in length, but its H projection changes from a point to the line ab.

The third position on the Plate shows the projections of the line when inclined to both H and V. The line has been revolved about the point A, so that its H projection makes an angle of 45° with V. The H projection in this case remains the same in length as in the second position, but its V projection now appears foreshortened.

Problem 1.—Draw the first position of the line as shown, and for the second position, revolve the line in a direction parallel to V about the point A, until it makes an angle of 60° with H, and find its H projection. For the third position, revolve the H projection of the second position about the same point, until it makes an angle of 30° with V.   *Time, 2 hours.*

Problem 2.—Draw the first position of the line as shown, and for the second position, revolve the line toward the right in a direction parallel to V about the point B, to an angle of 45° with H, and find its V projection. For the third position, revolve the H projection of the second position about the same point to an angle of 45° with V.   *Time, 2 hours.*

Problem 3.—Draw the first position of the line as shown, and for the second position, revolve the line about the point B, until its V projection is parallel to GL, and find its H projection. For the third position, revolve its H projection of the second position about point B, until it is perpendicular to GL.   *Time, 2 hours.*

PLATE 5.

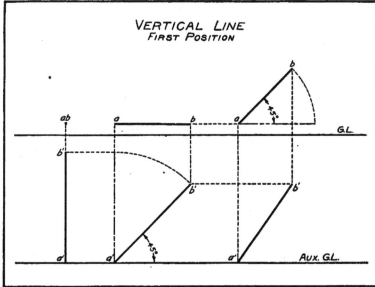

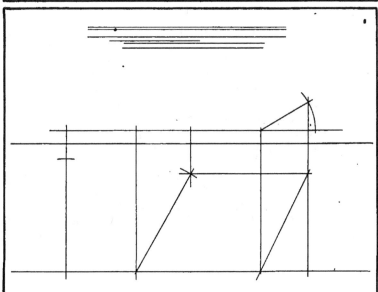

Layout for Problem 1.

PLATE 6

HORIZONTAL LINE
(First Position)

189. Example.

The first position on the Plate shows a line AB, 3-1/2 inches long. In its first position it is parallel to H and 2-3/4 inches from H, while the end A rests against the V plane. Since the line is parallel to H, its projection is shown as ab and its true length can be measured upon the H plane, and since its H projection is perpendicular to GL, hence perpendicular to V, its V projection is a point and is shown as $a'b'$.

The second position on the Plate shows the H projection of the line revolved parallel to H about the end A to an angle of 30° with V. Since the line remains parallel to H, its H projection remains of the same length as before, while its V projection is changed from a point to a line $a'b'$.

The third position on the Plate shows the V projection of the line revolved through an angle of 45° about the point A. The V projection in this case remains the same in length as in its second position, but its H projection now appears foreshortened.

Problem 1.—Draw the first position of the line as shown, and for the second position, revolve the line in a direction parallel to H about the point A, until it makes an angle of 45° with V. Then revolve its V projection about the same point until it makes an angle of 30° with H.   *Time, 2 hours.*

Problem 2.—Draw the first position of the line as shown, and for the second position, revolve the line toward the right about the point B in a direction parallel to H, through an angle of 45° with V. Then revolve its V projection about the same point through an angle of 30° with H.   *Time, 2 hours.*

Problem 3.—Draw the first position of the line as shown, and for the second position, revolve the line about the point B in a direction parallel to H, through an angle of 60° with V. Then revolve its V projection about the same point through an angle of 30° with H.   *Time, 2 hours.*

PLATE 6.

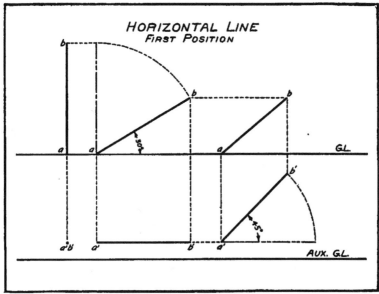

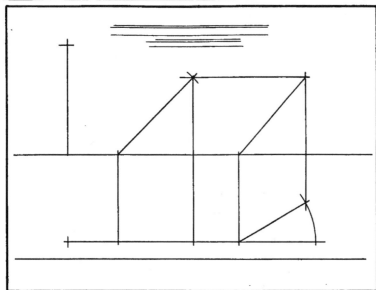

Layout for Problem 1.

PROJECTIONS OF PLANE FIGURES

190. Principles.

1. The representation of a plane figure or surface upon a plane of projection, may be either a line or a figure, depending on the relation of the figure to the plane.

2. The projection of a plane figure or surface parallel to a plane of projection, will be a similar figure, and have its true size shown upon that plane.

3. The projection of a plane figure or surface perpendicular to a plane of projection, will be a line.

4. The projection of a plane figure or surface inclined to a plane of projection, will be a foreshortened figure or surface.

The layout drawing shown on each of the following five plates, is for one of the problems dependent on the completed example. These layouts are intended to serve as a guide for analysis and method of procedure in working the problems.

PLATE 7

VERTICAL PLANE

191. Example. (Fig. A.)

A plane figure or surface $ABDC$ in space is 2 inches wide and
2-1/2 inches high. In its first position on the Plate the figure
is parallel to V, and its surface is 3/4 inch from V, and the edge
AC is 1/2 inch from H. Since the figure is parallel to V, its projection upon the V plane is a similar figure and is shown as
$a'b'd'c'$; also, since the figure is parallel to V, it is perpendicular to H; and its projection upon this plane is shown as $ab\ cd$.

The second position on the Plate shows the projections of
the figure when revolved in a counter-clockwise direction about
the edge AB, to an angle of 45°. Since the figure is now
inclined to V, the projection of its width is foreshortened and is
shown as $a'b'd'c'$. Its H projection, since the relation of the
edges AC and BD to the H plane remains unaltered, is shown of
the same length as before.

NOTE.—In the first position of the figure its height is measured on its V
projection, while its width may be measured either on V or on H. In the
second position of the figure the height is measured on V, while its width can
be measured only on H.

Problem 1.—Draw the first position of the figure as shown.
For the second position, draw projections of the figure revolved
about the edge CD through an angle of 30°.   *Time, 2 hours.*

Problem 2.—For the first position, draw the projections of the
figure when removed 1/4 inch farther from V. For the second
position, draw the projections of the figure when revolved about
the edge CD through an angle of 90°.   *Time, 2 hours.*

Problem 3.—Draw the projections of the figure, for its first
position, when its surface is 1-1/4 inches from V, and its edge
AC is 3/4 inch from H. For the second position, show the
projections of the figure revolved about its vertical center line
through an angle of 15°.   *Time, 2 hours.*

PLATE 7.

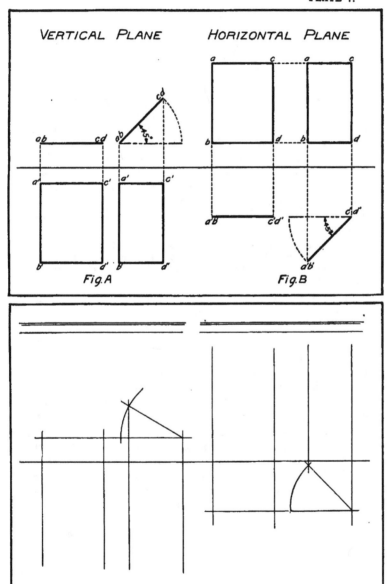

Layout for Problems 1, Fɪɢ. *A* and Fɪɢ. *B*.

PLATE 7

HORIZONTAL PLANE

192. Example. (Fig. B.)

A plane figure or surface *ABDC* in space is 2 inches wide and 2-1/2 inches long.   In its first position on the Plate the figure is parallel to H, while its surface is 1-1/2 inches from H, and the edge *BD* is 3/4 inch from V.   Since the figure is parallel to H, its projection upon H is a similar figure and is shown as *abdc*, while its V projection is a line and is shown as *a'b' c'd'*.

The second position shows the projection of the figure when revolved in a counter-clockwise direction about the edge *CD*, through an angle of 45°.   Since the figure is now inclined to H, the projection upon that plane is foreshortened and is shown as *abdc*, while the V projection remains unaltered in length from that shown in the first position.

NOTE.—In the first position the length of the figure is measured on H, while its width may be measured either on H or on V. In the second position the length of the figure is measured on H, while its width can be measured only on V.

Problem 1.—Draw the first position of the figure as shown. For the second position, draw the projections of the figure revolved about the edge *CD*, in a clockwise direction through an angle of 45°.   *Time, 2 hours.*

Problem 2.—For the first position, draw the projections of the figure when removed 1/2 inch farther from H.   For the second position, draw the projections of the figure when revolved about the edge *CD*, in a clockwise direction through an angle of 75°·   *Time, 2 hours.*

Problem 3.—Draw the projections of the figure, for its first position, when moved 1/4 inch farther from both H and V.   For the second position, draw the projections of the figure when revolved about a line perpendicular to V and passing through its center, through an angle of 15°.   Revolve the figure in either a clockwise or counter-clockwise direction.   *Time, 2 hours.*

PLATE 7.

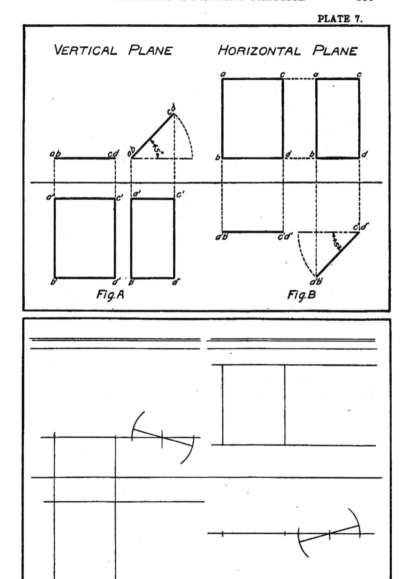

Layout for Problems 3, FIG. A and FIG. B.

PLATE 8

RECTANGULAR PLANE

193. Example.

A rectangular plane figure or surface *ABDC* is 2 inches wide and 3 inches long. In its first position the figure is parallel to H and its surface is 2 inches from H, while its edge *BD* is in contact with V. Since the figure is parallel to H, its true size can be measured upon the H plane and is shown by its projection *abdc*; also since it is parallel to H it is necessarily perpendicular to V, and is shown as *a'b' c'd'*, on the V plane.

The second position shows the projections of the figure when turned in a clockwise direction through an angle of 30° about the corner *D*, which remains in contact with V. Since the surface of the figure remains parallel to H, its H projection is of the same dimensions as in its first position.

The third position on the plate shows the figure turned from its second position, about the corner *D*, so that its V projection makes an angle of 30° with H. The V projection of the figure is not altered in length, but the H projection is foreshortened, and its true size can not be measured upon either H or V.

Draw the first position of the figure as shown in the example for any of the following problems:—

Problem 1.—For the second position, revolve H projection of the first position about the corner *b*, through an angle of 30° and find its V projection. For the third position, revolve the V projection of the second position about the corner *b'* through an angle of 30° in a clockwise direction.   *Time, 2-1/2 hours.*

Problem 2.—For the second position, revolve the H projection of the first position about the corner *d*, through an angle of 45°, and find its V projection. For the third position, revolve the V projection of the second position about *d'* through an angle of 90° in either direction.   *Time, 2-1/2 hours.*

Problem 3.—For the second position, draw the projections of the figure when revolved about its edge *CD* of the first position, through an angle of 30° in a clockwise direction. For the third position, revolve the H projection of the second position about the corner *d* through an angle of 45° in a clockwise direction.

Time, 2-1/2 hours.

PLATE 8.

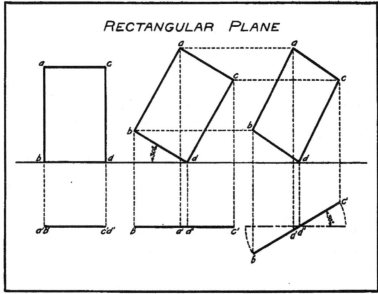

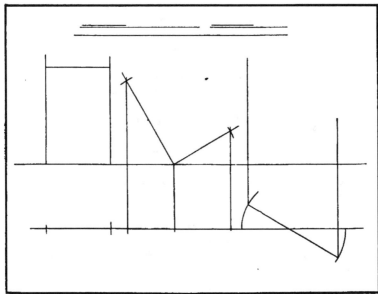

Layout for Problem 1.

PLATE 9

SQUARE PLANE

194. Example.

The plane shown on Plate 9 is a 2 inch square. In the first position the figure is shown parallel to H and 1–3/4 inches from the H plane. One corner is in contact with V and its edges make equal angles with the V plane. Both diagonals, *AC* and *BD*, are parallel to H, while *AC* is also parallel to V; the diagonal *BD* is perpendicular to V.

The second position shows the projections of the figure when its surface is revolved through an angle of 45° with H, about the diagonal *BD* as an axis. The diagonal *BD* remains parallel to H and perpendicular to V, while the diagonal *AC* remains parallel to V but is now inclined to H.

The third position shows the projection of the figure with both diagonals inclined to V, while *AC* is also inclined to H; the diagonal *DB* is parallel to H, as in the second position.

NOTE.—The true size of both diagonals of the figure may be measured in the H projection of the first position, while *AC* may also be measured in the V projection. In the second position the true length of *BD* can be found only on H, while *AC* is found on V. In the third position the true length of *BD* is found on H, while the true length of *AC*, since it is inclined to both planes, can not be measured on either plane of projection.

Draw the first position of the figure as shown in the example for the any of following problems:—

Problem 1.—For the second position, turn the V projection of the figure about *BD* as an axis through an angle of 30° in a counter-clockwise direction, and find the H projection. Then turn the H projection of the figure through an angle of 45° in a clockwise direction.   *Time, 2–1/2 hours.*

Problem 2.—For the second position, turn the V projection of the figure about the point *a'* through an angle of 15° in either direction, and find the H projection. Then turn the H projection of the figure until *dc* coincides with GL.   *Time, 2–1/2 hours.*

Problem 3.—For second position, turn the V projection of the figure about *BD* as an axis, through an angle of 30° in a counter-clockwise direction, and find the H projection. Then turn the H projection in a counter-clockwise direction through an angle of 45°.   *Time, 2–1/2 hours.*

PLATE 9.

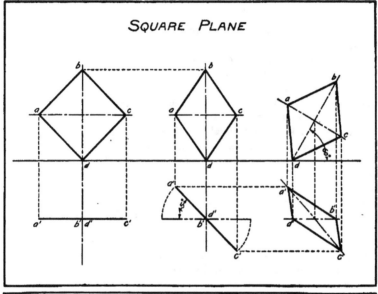

SQUARE PLANE

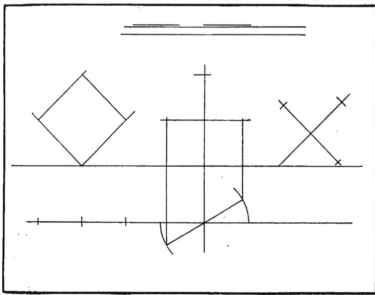

Layout for Problem 1.

PLATE 10

CIRCULAR PLANE

195. Example.

This plate shows the projections of a circle in three different positions with reference to the two planes of projection. The figure is 3 inches in diameter and in its first position is parallel to H and 1–1/2 inches from H, while its center is 1–3/4 inches from V. The second position shows the projections of the figure when turned through an angle of 45° in a counter-clockwise direction, about the diameter AB as an axis. The third position shows the figure when the diameter CD is inclined to both H and V.

NOTE.—The true length of AB is shown in the H projection of the three positions, since the line is parallel to the H plane. The true length of CD can be measured in H or V of the first position, also in the V plane of the second position.

Problem 1.—For the first position, draw a 2–3/4 inch circle, the dimensions for location remaining as in the example. For the second position, show the figure turned about AB as an axis, through an angle of 30° in either direction. For the third position, show the figure with the AB axis inclined to an angle of 75° with V, and parallel to H. *Time, 3 hours.*

Problem 2.—Draw the first position of the figure as shown. For the second position, show the figure turned through an angle of 30° about the point C in a counter-clockwise direction, the axis AB remaining parallel to H. For the third position, show the figure turned about B in a clockwise direction, until the axis AB makes an angle of 45° with V, and parallel to H. *Time, 3 hours.*

Problem 3.—Draw the first position of the figure, removed to a distance of 1–3/4 inches from H; the other dimensions remain as in the example. For the second position, show the figure turned about the axis AB through an angle of 60° in a clockwise direction. For the third position, show the figure when the H projection of the second position is turned about its center in a clockwise direction through an angle of 90°. *Time, 3 hours.*

PLATE 10.

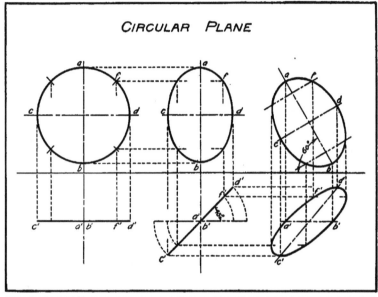

CIRCULAR PLANE

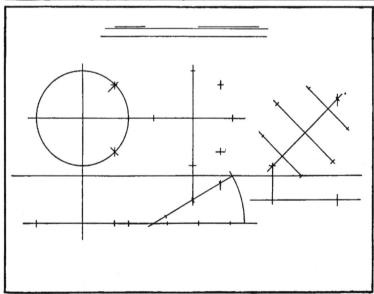

Layout for Problem 2.

PROJECTIONS OF SOLIDS

196. The principles given for the projection of lines and planes are also embodied in the projections of solids. Since a surface is a combination of lines which meet at projected points, so is a solid a combination of surfaces which meet in projected lines. It is therefore only necessary to project the bounding lines of the surfaces of a solid to obtain its projection.

197. Principles.

1. The top and front views of any point on a line bounding a surface must lie on a line drawn perpendicular to a horizontal line, called a **Ground Line,** or **G L.**

2. The front and end views of any point on a line bounding a surface must lie on a line drawn perpendicular to a vertical line, called the **Vertical Axis,** or **V G L.**

3. In third angle projection, the top view or plan of any point, line, surface or solid must be placed above the front view or elevation.

4. In third angle projection, the view showing the left side of a point, line, surface or solid must be placed at the left of the front or top views, while a view showing the right side must be placed at the right of the front or top views.

5. An end view is usually placed at the end of the front view, although occasions may arise when it is desirable to place it opposite the top view.

Before beginning any exercise the student should carefully study the layout drawings shown above the completed example.

PLATE 11

RECTANGULAR PRISM

198. Example.

This plate shows the method employed in representing a solid object in various positions on two planes of projection. The first position shows the top and front views of a prism, 3/4 inch wide, 1-1/4 inches long and 3-1/2 inches high, standing vertically on an auxiliary horizontal plane, with its face *ABDC* in contact with V. The distance between the auxiliary H plane, and the H plane is 4 inches. The second position shows the top view and front view of the prism when turned toward the left about a long edge of its base, to an angle of 45° with the auxiliary H plane, the face *ABDC* being in contact with V. The third position shows the top view and front views when inclined to both H and V.

The method of procedure for drawing an object as represented in the third position is as follows: First draw the top and front views as shown in the first position. For the second position, draw the front view of the same dimensions as shown in the first position and inclined to any desired angle with H, and project the top view. For the third position, draw the top view as shown in the second position and inclined to any angle with V, and project the front view.

Problem 1.—Draw the top and front views of a prism with a 1-1/2 inch square base and 3 inch altitude, in the three positions shown in the example. *Time, 2 hours.*

Problem 2.—Draw a prism having a 1-1/2 inch square base and 3-1/4 inch altitude, in the first position, as in the example. For the second position, turn the prism toward the right, through an angle of 60° about an edge perpendicular to V. For the third position, turn the H projection of the prism to the right through an angle of 30°, and project the front view. *Time, 2-1/2 hours.*

Problem 3.—With the dimensions as given for the example, draw, for the first position, the top and front views of a vertical prism having its wide faces parallel to V, and showing the prism removed 1/2 inch from V. For the second position, turn the prism in a clockwise direction and parallel to V about a base edge through an angle of 30°. For the third position, revolve the top view of the second position in a clockwise direction through an angle of 60°, and project the front view. *Time, 2-1/2 hours.*

PLATE 11.

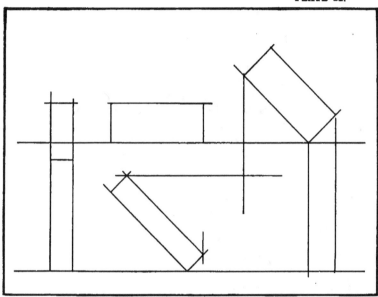

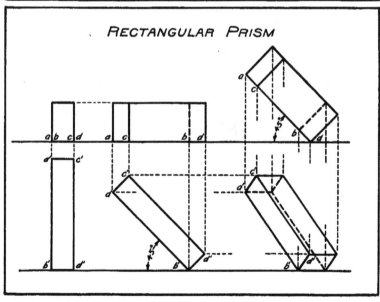

RECTANGULAR PRISM

PLATE 12

TRIANGULAR PRISMS

199. Example.

This plate shows the projections of two triangular prisms with a surface of one resting on an arris of the other. The first position shows the horizontal prism resting on an auxiliary H plane, with its ends parallel to the V plane, while the inclined prism is shown with its ends perpendicular to V. The second position shows the top and front views when the H projection of the first position is turned through an angle of 30° with V.

NOTE.—Both prisms have 2 inch faces, and the altitude of a 2 inch triangle may be shown as being equal to the $\sqrt{3}$.

Problem 1.—Draw the front and top views of a horizontal triangular prism as shown. Then draw the inclined prism making the angle of inclination 45° with H. Then move the top view to the right and incline it to an angle of 30° to GL, and project the front view.   *Time, 2-1/2 hours.*

Problem 2.—Draw a rectangular prism 1-1/2 × 1-3/4 × 4 inches with one of its wide faces resting on a triangular prism as shown in the example, the rectangular prism making an angle of 30° with H. Then move the top view to the right and turn it through an angle making 45° with GL, and project the front view.   *Time, 2-1/2 hours.*

Problem 3.—Move the horizontal prism, as shown in the example, to the right, and place the inclined prism on the left, making an angle of 45° with H, and project the top view. Then move the top view toward the right and incline it to an angle of 45° with GL, and project the front view.   *Time, 2-1/2 hours.*

PLATE 12.

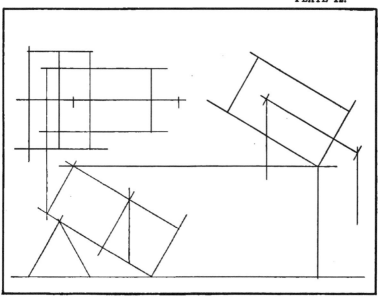

TRIANGULAR PRISMS

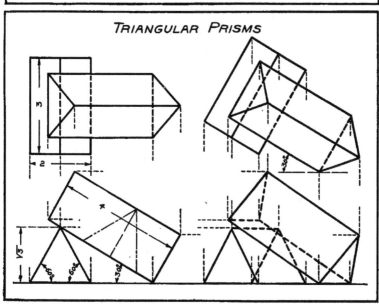

SECTION III

WORKING DRAWINGS

200. Drawings intended as working drawings are made in accordance with the principles of orthographic projection, called, for short, projection. Although the experienced drafts-man rarely gives any thought to the term orthographic projection in the execution of his drawing, he must nevertheless be familiar with the art of projection. Orthographic projection is the theoretical side of a draftsman's work; making a working drawing is its practical application.

Since the function of a working drawing is to convey definite information to a workman to make or construct the object represented, it must contain as many views of the object as are required for a clear interpretation of what the drawing is to represent.

201. A working drawing should contain all the dimensions necessary for the construction of the object represented. Dimensions given in one view need not, as a rule, be repeated in other views.

All lines, dimensions, and explanatory notes should be so clearly drawn as to make a misinterpretation of the drawing impossible.

202. In making a working drawing, the principles of orthographic projection are occasionally violated, if the violation will give more clearness to the legibility of the drawing. There are, however, no fixed rules for violations of projection. The draftsman of experience can judge, and will decide when violations are desirable. For examples, see Plates 31 and 37.

It is not customary in making working drawings to make sectional views of bolts, shafting, or the arms of pulleys, etc. For illustration, the section of the pulley on Plate 37 is taken vertically and a small portion of the arm at the rim and hub is cut by the vertical cutting plane. These small portions, in accurate projection drawing, would be cross-sectioned, but in practice such parts are so familiar to the workman that such minute detail is omitted. Similarly, in drawing a bolt or nut, or a round shaft, etc., only one view is necessary, as the other views are so well known and understood that the time spent drawing them would, in practice, be considered wasted.

PLATE 13

RECTANGULAR PRISM

(Projections upon the Horizontal and Vertical Planes)

203. Example.

This plate shows the top and front views of a rectangular prism or solid with reference to two planes of projection. In its first position on the plate the prism has one of its surfaces in contact with V. In the second and third positions it is turned through the angles indicated, one corner remaining in contact with V.

Purpose.—The purpose to this plate is to show the method of projection of such simple objects as a brick, rectangular tile, box, etc. Such objects consist of six rectangular planes which are mutually at right angles to each other. All details with regard to form and dimensions can be shown by a top and a front view.

Problem 1.—Draw the top and front views of a prism having a 1-1/2 inch square base and 3 inch altitude, in positions similar to those shown in the example.   *Time*, 2-1/2 *hours*.

Problem 2.—Draw the top and front views of a solid having a 1-1/4 × 3/4 inch base and 2-3/4 inch altitude. For the first position, let one of the narrow surfaces be in contact with V. In the second position, turn it in a counter-clockwise direction through an angle of 30° with V. Next turn it in a clockwise direction through an angle of 75° with V.   *Time*, 2-1/2 *hours*.

Problem 3.—Draw a solid of the dimensions shown in the example. For its first position, let it be removed 3/4 inch from V and 1-1/4 inches from H. In the second position, show it turned about its vertical axis, in a clockwise direction, through an angle of 30°. In the third position, show it turned through an angle of 45°, about its vertical axis in a counter-clockwise direction.   *Time*, 2-3/4 *hours*.

Optional Practice.—These problems afford an opportunity for the application of shade lines, also line shading. See articles on these subjects.

Optional Practice.—Calculate the surface, volume, and weight; if made of cast iron, of the object drawn. (See Art. 339.)

Cast iron weighs .26 pound per cubic inch.

PLATE 13.

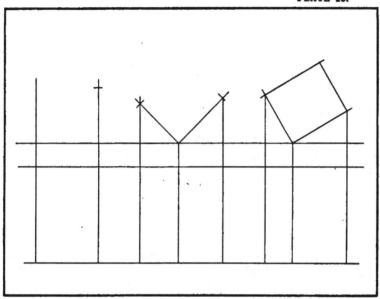

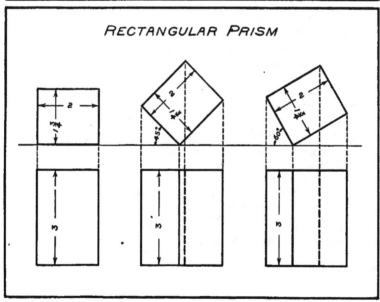

RECTANGULAR PRISM

PLATE 14

TRIANGULAR AND HEXAGONAL PRISMS

(Projections on Horizontal, Vertical and Profile Planes)

204. Example.

This drawing illustrates the manner of showing the top, front and end views, of two prisms placed in definite relation to the three planes of projection.

The prisms have one corner in contact with V and both are removed 5/8 inch from the profile plane, P. The top face of the triangular prism is 1/2 inch from H, while the top face of the hexagonal prism is 3/4 inch from H.

Purpose.—This plate presents a study of solids such as hexagonal plinths, columns, etc., whose surfaces are bounded by straight lines, some of which are parallel to V and also to P, while others are parallel to H and P.

Problem 1.—Draw three views of both prisms of the dimensions as shown in the example, each having one face in contact with V. Let one corner of each prism be 1 inch from P, also let the top surface of both be 3/4 inch from H. *Time, 3-1/2 hours.*

Problem 2.—Draw three views of a triangular prism each surface of which is 3-3/4 inches altitude and 2-1/4 inches wide, the prism being removed 1/2 inch from V and the top face 1 inch from H; let two of its faces make equal angles with V. Draw three views of an hexagonal prism, each face of which has 3-3/4 inches altitude and 1 inch width, assuming the center of the top view to be as shown in the example. *Time, 3-1/2 hours.*

Problem 3.—Draw three views of both prisms revolved through an angle of 15° about their axes in a clockwise direction, of dimensions and locations as shown in the example. *Time, 4 hours.*

NOTE.—For the method of drawing the plan views, refer to GEOMETRICAL PROBLEMS (Arts. 259 and 262).

Optional Practice.—These problems afford an opportunity for the application of shade lines, also line shading.

Optional Practice.—Calculate the surfaces, volumes; and weights, if made cast steel, of the objects drawn. (See Arts. 340 and 341.)

Cast steel weighs .28 pound per cubic inch.

PLATE 14.

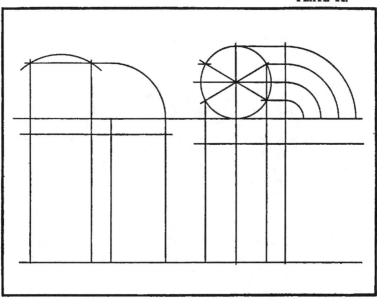

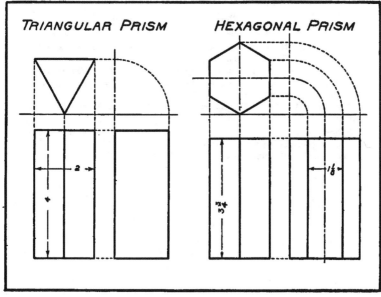

TRIANGULAR PRISM

HEXAGONAL PRISM

PLATE 15

TRIANGULAR AND PENTAGONAL PYRAMIDS

(Projections upon Three Planes)

205. Example.

This drawing shows three views of two pyramids resting on an auxiliary H plane, which is removed 4-3/4 inches from H. One side of the base of each pyramid is in contact with V. The axis of the triangular pyramid is 1-3/4 inches from P, while the axis of the pentagonal pyramid is 1-3/8 inches from P.

Purpose.—This plate presents a study of two solids, the surfaces of which are bounded by straight lines, most of which are not parallel to any two planes of projection.

Problem 1.—Make a drawing showing three views of the pyramids as shown in the example, but removed 1/2 inch from V, and their nearest corner 1 inch from P.   *Time, 3-3/4 hours.*

Problem 2.—Make a drawing showing three views of the pyramids as shown in the example, with one side of the base parallel to P, and one corner in contact with V.   *Time, 4 hours.*

Problem 3.—Show three views of each pyramid as shown in the example, when two of their faces make equal angles with V. Assume the points of location.   *Time, 4 hours.*

NOTE 1.—For the method of drawing the top views, refer to GEOMETRICAL PROBLEMS (Arts. 259 and 261).

NOTE 2.—The true lengths of the edges of these pyramids, whose projections are perpendicular to GL, are shown in the end views, see PROJECTION UPON THREE PLANES (Art. 166).

NOTE 3.—The true length of the slant height, also the arris, must be measured on P.

Optional Practice.—Calculate the surfaces, volumes, and weights, if made of brass, of the objects drawn. (See Arts. 307, 343 and 345.)

Brass weighs .30 pound per cubic inch.

PLATE 15.

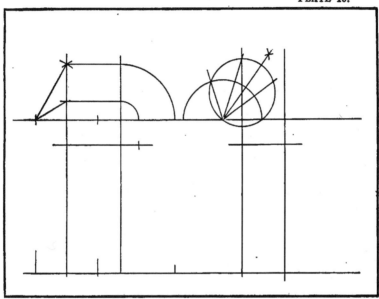

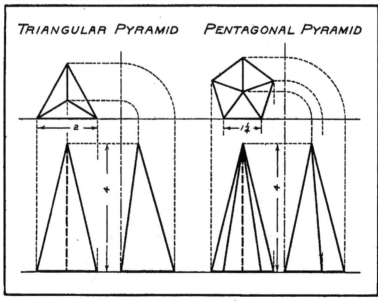

TRIANGULAR PYRAMID   PENTAGONAL PYRAMID

PLATE 16

MORTISE AND TENON JOINTS

206. Example.

This drawing shows the front and top views of two mortise and tenon joints. The one joint consists of two parts, the other of three parts.

Mortise and tenon joints find extensive applications in many kinds of wood constructions, such as tables, chairs, doors, house building, etc.

Purpose.—The object of this plate is to show how simple objects consisting of two or three parts can be shown; also how the internal construction of such similar objects is shown by the aid of hidden lines.

In general only two views are required for objects of this kind, although Problems 2 and 3 require end views to be drawn, to cultivate the student's powers of visualizing and developing a third view, when two views are given.

Problem 1.—Make a drawing showing the front views of both joints, with the dimensions as in the example, revolved through an angle of 180° in the plane of the paper, and project the top views.   *Time, 2-1/2 hours.*

Problem 2.—Make a drawing showing the front, top and end views of the joint shown on the left of the example. Assume the places for location, and show the end view on the right of the front view.   *Time, 2-1/2 hours.*

Problem 3.—Make a drawing showing the front, top and end views of the joint shown on the right of the example. Assume the places for location, and show the end view on the right of the front view.   *Time, 3 hours.*

NOTE 1.—All lines entering into the execution of these drawings, with the exception of two, are vertical or horizontal.

NOTE 2.—The GL. denoting the position of the objects with reference to the two planes of projection, has been omitted in the plate. The student should now be able to make a simple projection without reference to the GL. He will also have had practice enough to know that the size or shape of an object is not influenced by its position in space with reference to imaginary planes.

Optional Practice.—Make a bill of material showing the thickness, width and length of the pieces required, to construct the object drawn.

PLATE 16.

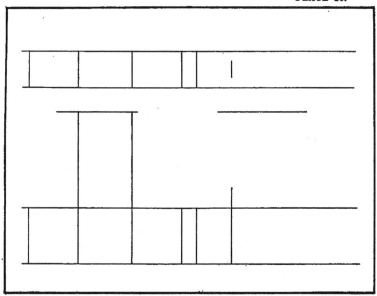

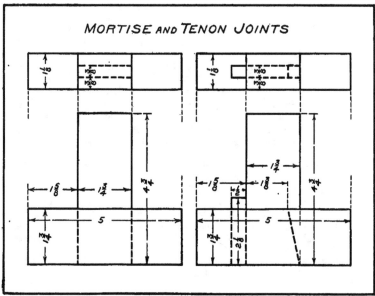

MORTISE AND TENON JOINTS

PLATE 17

FLAT LINK

207. Example.

This plate shows two views of a flat link. Links of this kind are made of metal, usually of steel or wrought iron; they find extensive application in machine and building construction.

Purpose.—The object of the problems given in connection with this plate is to afford practice in drawing circles, also circular arcs, which are tangent to circular arcs and straight lines.

Problem 1.—Show the edge and top views of a link making both ends similar to the right-hand end, as shown in the example.

Time, 2–1/2 hours.

Problem 2.—Show the edge and top views of a link having the inside and outside diameters of both ends increased 1/2 inch.

Time, 3 hours.

Problem 3.—Draw three views of a link increasing every dimension 1/4 inch. Place the edge and top views as shown in the example, and the end view opposite the right-hand end of the top view.

Time, 3–1/2 hours.

NOTE 1.—For drawing the small arcs see problems relating to LINE AND TANGENT ARCS (Art. 242).

NOTE 2.—The small circular arcs, when placed as shown in the example, are technically called *fillets*. They are used to avoid a sharp corner, which in many cases is objectionable in machine and building construction.

Optional Practice.—These problems afford an opportunity for the application of shade lines for both views. The edge view may also be line shaded.

Optional Practice.—Calculate the weight of the object drawn, if made of wrought iron, which weighs .28 pound per cubic inch. (See Arts. 324 and 339.)

NOTE.—The weight of the link shown in the example, may be found by first finding the volume in cubic inches, then multiplying this by the weight per cubic inch of the material.

The following method for finding the volume is approximate, but is accurate enough for practical purposes. The area of the figure will be approximately the area of the annulus on the left, plus the area of the annulus on the right, plus the area of the bar. This sum multiplied by the thickness will give the volume.

$$\text{Volume} = 1/2[.7854(3 + 1\text{-}3/4)(3 - 1\text{-}3/4) + .7854(2 + 3/4)(2 - 3/4) + 1\text{-}1/4 \times 4\text{-}3/8]$$

PLATE 17.

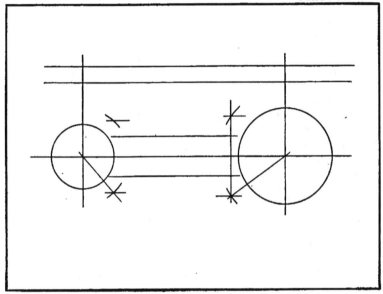

FLAT LINK

PLATE 18

WALL BRACKET

208. Example.

The drawing on this plate shows the side, top and right-hand end views of a small wall bracket, such as may be employed for fastening a shelf or platform to a vertical surface or wall of a building. Brackets of this design are made of wood.

Purpose.—The purpose of this plate is to show how a simple object consisting of three pieces can be represented. The drawing, also, is intended to show how the internal construction of a mortise and tenon joint can be shown by hidden lines or by a detail section. It furthermore shows how the dimensions can be effectively distributed over the drawing.

Problem 1.—Make a drawing showing the front and top views of the bracket as shown in the example; also show a left-hand view, instead of the right-hand end view, as illustrated.

Time, 3-1/2 hours.

Problem 2.—Make a drawing of the side view of the bracket shown in the example; also views looking from the under side and left-hand end. *Time, 3-1/2 hours.*

Problem 3.—Draw three views of a bracket, similar to that shown in the example. The bracket to be constructed of stock 1-3/8 inches square. The height and length of the bracket are to be 6 inches and 8 inches respectively, and the dimensions of the diagonal brace are to be increased from 3-3/4 inches to 4 inches. The other changes necessary are at the option of the student. *Time, 3-3/4 hours.*

Note.—For the cross-hatched section see articles on CROSS-HATCHING and SECTIONS.

Optional Practice.—Make a bill of material showing the thickness, width, and length of each piece required, for the object drawn.

PLATE 18.

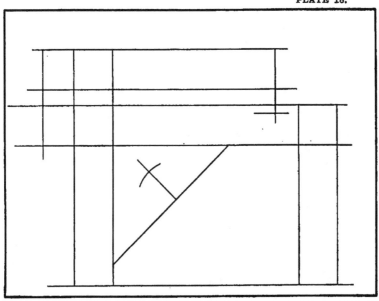

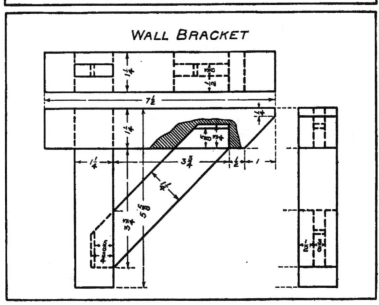

WALL BRACKET

PLATE 19

WALL BRACKET DETAILS

209. Example.

This is a drawing showing the details of the preceding example. It shows two views of the top member with a revolved section, and three views of each of the remaining two members.

Purpose.—This plate illustrates a suitable arrangement of views for the details of a simple object similar to a wall bracket. It furthermore shows how an end view may be shown directly on a front or side view, in the form of a revolved section. In Problem 3, the student will note how the character of some lines may change from full to hidden, or from hidden to full, dependent on the view chosen.

Problem 1.—Make a drawing showing the end views of the two horizontal details, on the right, instead of on the left; also place the vertical detail to the right of the horizontal details.

Time, 3–3/4 hours.

Problem 2.—Make a detail drawing of Problem 3, Plate 18. Place the views similar to those of the example. *Time, 4 hours.*

Problem 3.—Make a drawing of the details shown in the example, with the views of each detail interchanged, in any way which will not violate the rules of correct projection. The arrangement for location is at the option of the student.

Time, 4–1/2 hours.

Note 1.—For the cross-hatched surface of the left-hand detail, see article on Revolved Section (Art. 171).

Note 2.—When the views of a detail, in Problem 3, are interchanged, the character of some lines will also change.

PLATE 19.

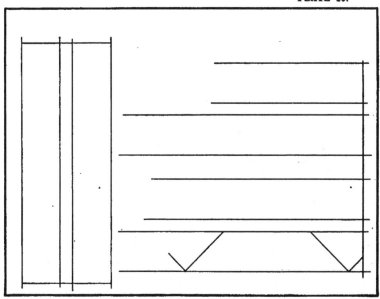

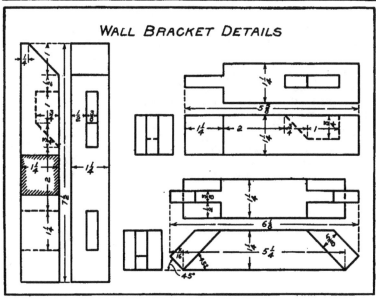

WALL BRACKET DETAILS

PLATE 20

EYE BOLTS AND HOOK

210. Example.

This plate illustrates the top and front views of metal fastenings, such as are used for securing ropes, cables, etc. The eye bolt on the left may be made of cast iron or brass, but is usually made of steel by drop forging. The one shown in the center of the plate is a forging made of wrought iron or steel. The hook shown at the right is generally a drop forging made of iron or steel, but it may also be a steel casting.

Purpose.—This plate affords practice in the study of circles and circular arcs tangent to straight lines. It is typical of objects where a revolved section may be used to show the shape of the cross-section, as shown in the front view of the hook.

Problem 1.—Draw the front views of the fastening shown in the example, and project views looking from below the front views, instead of from the top.   *Time*, 4 *hours*.

Problem 2.—Draw the front views of any two fastenings as shown in the example, and project views looking from below and from the right.   *Time*, 4–1/2 *hours*.

Problem 3.—Multiply every dimension of the two fastenings shown on the right of the plate by 1–1/4, and draw three views of each.   *Time*, 5 *hours*.

NOTE.—For the cross-hatched surface, see REVOLVED SECTION (Art. 123). For tangent arcs, see LINES AND TANGENT ARCS (Art. 246).

Optional Practice.—Calculate the volume and weight, if made of steel, of the hook shown in the example. (See Arts. 337 and 342.)

NOTE.—First calculate the volume of the cylinder, which is 7/8″ diameter and 2–1/2″ long; next calculate the volume of the prism, which is 7/8″ square and 1″ long, and last calculate the volume of a semi-cylinder, which is 7/8″ high and 2″ diameter. The sum of the three calculations will give the total volume. The volume multiplied by .28 will give the weight.

PLATE 20.

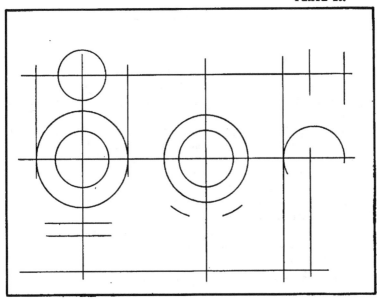

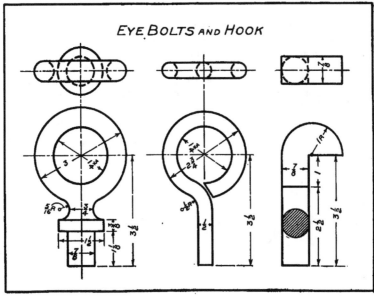

EYE BOLTS AND HOOK

PLATE 21

SMALL BENCH

211. Example.

This plate shows the front, top, and end views of a small bench. A small bench, as shown, is usually made of some hard wood as oak or ash, but may also be made of soft woods, such as poplar or white pine.

Purpose.—The purpose of this plate is to illustrate how a large object can be drawn to a reduced scale. It also shows how a portion of a surface may be broken away to show more clearly the construction underneath. In complicated pieces of machinery or wood construction, a surface shown broken on a drawing will change hidden lines into full lines, thereby making the details more easily understood.

Problem 1.—Draw the front and end views of the bench as shown in the example, and a view looking from the under side, instead of the top view.   *Time, 4 hours.*

Problem 2.—Make a detail drawing showing the necessary views of the top, one rail, and one leg. Draw to a scale of half size.   *Time, 4–1/2 hours.*

Problem 3.—Draw three views, placed as in the example, of a bench which is to be 14 inches wide, 28 inches long, and 12 inches high, all other dimensions remaining as shown.
                    Time, 4–1/2 hours.

Optional Practice.—Make a bill of material showing the dimensions of the pieces required; also calculate the number of square feet of stock, for the object drawn.

PLATE 21.

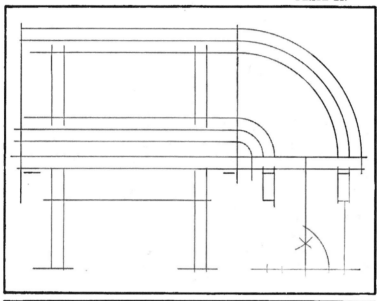

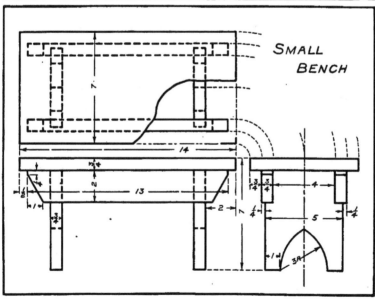

SMALL BENCH

PLATE 22

NUT WRENCH

212. Example.

This drawing shows two views of a wrench, such as is used for turning bolts, nuts, etc. Instruments of this kind are always made of steel.

Purpose.—The object of this plate is to afford practice in reading lines which are not horizontal or vertical, also in reading circular arcs whose centers are on lines which are inclined to a horizontal line. The problems given will afford practice in joining a number of circular arcs to produce a smooth curve.

Problem 1.—Make a drawing of the top view of the wrench, shown in the example, revolved through an angle of 180° in the plane of the paper, and project the front view. Show a revolved section in the handle.   *Time, 3–1/2 hours.*

Problem 2.—Make a drawing of a double-ended wrench, showing both ends like the right-hand end in the example, the width of the handle to be 5/16 inch thick and 1–1/4 inches wide throughout its length. Show a revolved section in the handle.
Time, 3–1/2 hours.

Problem 3.—Draw a top and a front view of a nut wrench having all the dimensions, as given in the example, multiplied by 1–1/2.   *Time, 4 hours.*

NOTE.—The length of this drawing must be contracted, to confine it to the limits of the paper, by breaking out a portion in the handle. For illustrations, see Fig. 57. It will be necessary for the student to lay out the full length of the handle on a separate piece of paper to obtain the proper angularity of the lines.

PLATE 22.

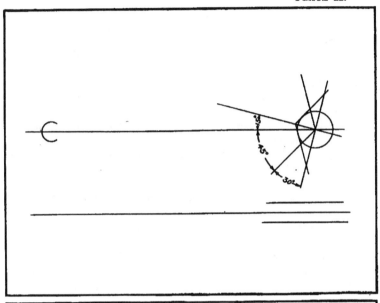

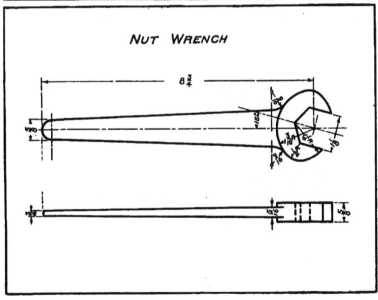

NUT WRENCH

PLATE 23

BRACKET SHELF

213. Example.

This plate shows three views of a small bracket shelf, used for holding ornaments, as a small clock, candle-sticks, etc. Brackets of this kind may be made of any suitable wood.

The top view, shown on the front view, is taken above the shelf, which necessitates cross-sectioning the back and showing the bracket underneath in hidden lines.

Purpose.—The object of this plate is to show an illustration of a number of circular arcs which meet but are not tangent to each other. It also shows how a top or plan view may be placed directly on the front view, thereby economizing in the size of the sheet of paper required.

Problem 1.—Make a drawing of the bracket as shown in the example, with the thickness of each of the three pieces entering into its construction increased by 1/8 inch. All other dimensions to be as shown.   *Time, 3–3/4 hours.*

Problem 2.—Make a drawing of the front and side views of the bracket as shown in the example, and show the top view placed on the side view.   *Time, 3–3/4 hours.*

NOTE.—Place the top view between the shelf and the upper end of the back.

Problem 3.—Make a drawing of a bracket shelf which is to be 15 inches long, 4 inches wide, and have the upper surface of the shelf 4–1/2 inches from the lower end of the back. All other dimensions, with the exception of the circular arcs, are to be as shown in the example.   *Time, 4 hours.*

NOTE 1.—The change in the dimensions of length will require a change in the radii of the circular arcs for the bracket. The design of this is left to the student.

NOTE 2.—Both front and side views must be constricted to reduce the length to the limits of the paper, by breaking out a piece. For illustrations, see Fig. 57. Place the top view in the broken out space of the front view.

Optional Practice.—Make a bill of material showing the dimensions of the pieces required, for the object drawn.

PLATE 23.

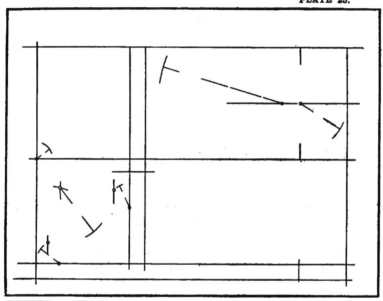

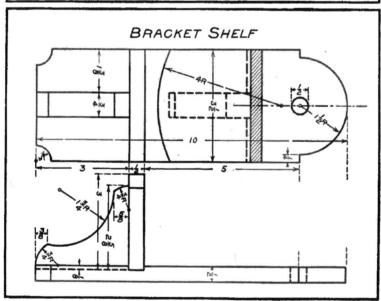

BRACKET SHELF

PLATE 24

214. Example.

This drawing shows two views of each of the three details of the small bracket shown in Plate 23.

Purpose.—This plate is intended to illustrate one arrangement for a suitable distribution of the three pieces required for the object.

Problem 1.—Make a detail drawing showing two views of each piece for the Bracket Shelf in Problem 1, Plate 23.

Time, 3–3/4 hours.

Problem 2.—Make a detail drawing showing three views of each piece for the Bracket Shelf in Problem 2, Plate 23. Arrange the views in some suitable order which is different from that shown in the example. *Time, 4 hours.*

Problem 3.—Make a drawing showing two views of each piece for the Bracket Shelf in Problem 3, Plate 23, and show revolved sections in the front views of the back and shelf. Arrange the views in some order different from that shown in the example.

Time, 4 hours.

PLATE 24.

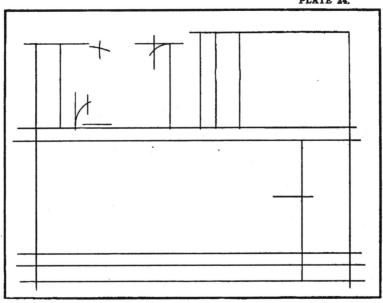

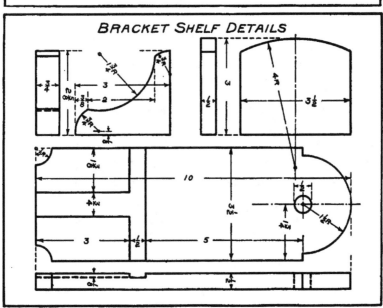

BRACKET SHELF DETAILS

PLATE 25

BELL CRANK

215. Example.

This plate shows two views of a cast-iron bell crank. This or similar devices find extensive applications in various kinds of machines. The function of the bell crank shown in the example is to translate a horizontal reciprocating motion into a vertical, or a vertical to a horizontal motion.

Purpose.—The object of the problems depending on this plate is to give practice in drawing circular arcs and circles, and lines which are not horizontal or vertical. The problems also give practice in finding a number of hidden lines, and also in dimensioning.

Problem 1.—Make a drawing of the front view of the bell crank as shown in the example, and place the end view on the left of the front view. Show a revolved section in each arm, assuming the arms to be rectangular in cross-section. *Time, 3–1/2 hours.*

Problem 2.—Make a drawing showing the front view of the bell crank shown in the example, revolved through an angle of 90° in a clockwise direction, and project the right-hand end view. Show a revolved section in each arm, assuming the arms to be flat but having semi-circular edges. *Time, 4 hours.*

Problem 3.—Make a drawing showing the front, end, and top views of a bell crank with every dimension shown in the example multiplied by 1–1/2. *Time, 4–1/2 hours.*

NOTE.—Since this drawing will be too large for the paper, it will be necessary to break both arms and show them contracted in length. It will be well for the student to make a full-size sketch of the front view on a separate piece of paper, to obtain the correct angularity of the lines.

Optional Practice.—Find the length of the circular arc described, by the center of the hole at the end of the vertical arm, when the bell crank drawn, is turned through an angle of 10 degrees, about the center of the 2 inch hub. (See Art. 331.)

Find the number of degrees which the horizontal arm of the bell crank describes, when the center of the hole moves through an arc of 3 inches, about the center of the 2 inch hub. (See Arts. 302 and 331.)

PLATE 25.

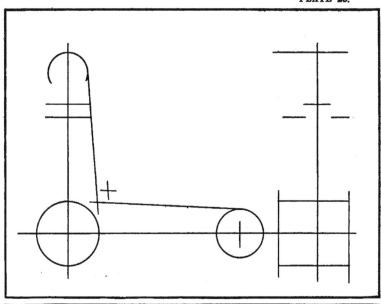

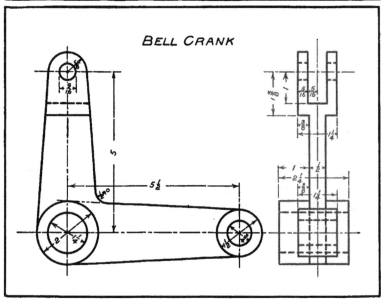

BELL CRANK

PLATE 26

TOWEL ROLLER

216. Example.

This drawing shows three views of a simple towel roller, which is to be made of hard wood, preferably maple, as this wood will not readily absorb moisture.   The bracket at the left is provided with a round hole and the bracket on the right is provided with a slot, to receive the roller.

Purpose.—The object of the problems given in connection with this plate, is to familiarize the student with the conventional method of showing long pieces of uniform cross-section in a comparatively short space.   It will be seen by referring to the example, that the break in the roller shows a round cross-section. The problems furthermore give practice in connecting two parallel lines by two tangent arcs, as shown in the front views of the back board.

Problem 1.—Draw a front view, a left-hand end view, and a view looking from below, instead of the top view, of the towel roller shown in the example.   *Time, 4–1/2 hours.*

Problem 2.—Draw three views of the towel roller, shown in the example, with the roller lowered 1/2 inch below its present position.

Lowering the roll will necessitate a new design for the end view of the bracket.   The design is at the option of the student.
                                                             Time, 5 hours.

Problem 3.—Make a detail drawing showing two views of each piece of the towel roller as shown in the example.
                                                             Time, 5 hours.

NOTE.—For connecting two parallel lines by tangent arcs, see LINES AND TANGENT ARCS (Art. 250).

Optional Practice.—These problems afford an opportunity for shade lines, also line shading. See articles on these subjects.

Optional Practice.—Make a bill of material showing the sizes of stock required to construct the object drawn.

PLATE 26.

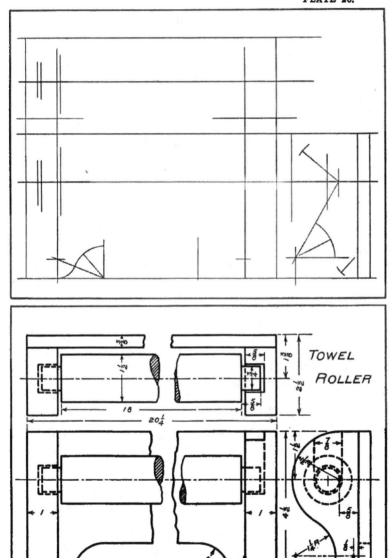

TOWEL
ROLLER

PLATE 27

HOLLOW CYLINDERS

217. Example.

This plate shows the top and front views of three hollow cylinders resting on an *auxiliary* H plane, which is removed 4-1/2 inches from H. Their vertical axes are removed 1-1/2 inches from V.

Cylinders similar to those shown may be made of wood or of metal, depending upon the functions which they are to perform. The cylinder shown on the left is a plain hollow cylinder, or sleeve. The cylinder in the center is called a hollow cylinder with two round flanges. The one on the right is a hollow cylinder of two diameters and one square flange.

Purpose.—The object of this plate is to illustrate three methods of showing the internal construction of objects of this character: the first by the sole use of hidden lines; the second by a part or detail section; the third by a full-length half-section. The problems will afford practice in drawing concentric circles in both full and hidden lines, in cross-sectioning, and in placing an object in a definite relation to two planes.

Problem 1.—Draw two views of each cylinder shown in the example. Change the outside surface of the first into a 2-1/4 inch square; change the outside of the second into hexagons, and change the outside of the third into octagons. Let the diameters given in the example, be the diameters of inscribed circles for the required polygons. Show each in half cross-section.

Time, 4 hours.

NOTE.—For drawing top views see GEOMETRICAL PROBLEMS (Arts. 260, 262 and 264).

Problem 2.—Draw two views of each cylinder shown in the example, having their axes horizontal. Let the axes be 1-3/4 inches from H and perpendicular to V, and let the ends of the cylinders be removed 1/2 inch from V. Show each cylinder in half full-length cross-section.   *Time, 4-1/2 hours.*

Problem 3.—Draw two views of three cylinders, similar to those shown in the example, in a horizontal position. Let the axes be 1-7/8 inches from H and perpendicular to V, and let the ends of the cylinders be 1/4 inch from V. For dimensions of the cylinders multiply each dimension shown in the example by 1.25.   *Time, 4-1/2 hours.*

PLATE 27.

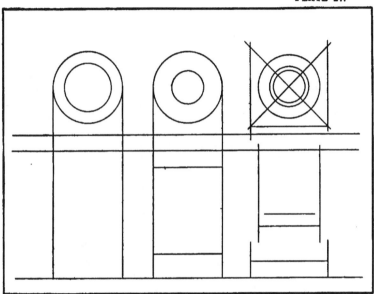

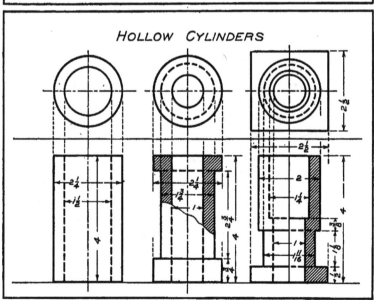

HOLLOW CYLINDERS

PLATE 28

PANEL AND SASH JOINTS

218. Example.

The two joints shown on this plate find application in making doors and window sashes. The joint shown on the left has a groove cut into the stile and rail to hold a panel, while the joint on the right has a rebate in both stile and rail to hold a pane of glass and putty.

They are both mortise and tenon joints, but are more complex than those shown on Plate 16.

Purpose.—The purpose of this plate is to show two objects which require considerable thought in the correct placing of hidden lines. Two views, one a front view, are all that are required for the construction of either joint. The third view is given to afford drill in a complete analysis of the mortises and tenons, also in the hidden lines.

Problem 1.—Make a drawing showing front, end, and top views of both joints shown in the example, but having the rail or horizontal member withdrawn 1/2 inch from the stile or vertical member.   *Time, 4 hours.*

Problem 2.—Make a drawing of the front views of both joints as shown in the example and project the bottom and right-hand views instead of the top and left-hand views.

Time, 4-1/2 hours.

Problem 3.—Make a drawing showing the front, end, and top views of two joints, similar to those shown in the example, with the following specifications:

Rail, 1-1/8″ × 2″ × 4″.
Stile, 1-1/8″ × 1-3/4″ × 5″.
Groove and rebate, each 3/8″ wide, 1/2″ deep.
Mortise and tenon, 1″ wide.

NOTE.—Show a revolved section in the stile of both joints. The details of design are to be supplied by the student.

Time, 5 hours.

Optional Practice.—Make a bill of material showing the thickness, width, and length of the pieces required, to construct the object drawn.

PLATE 28.

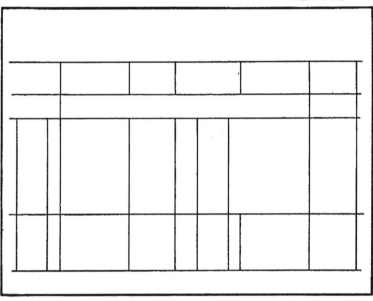

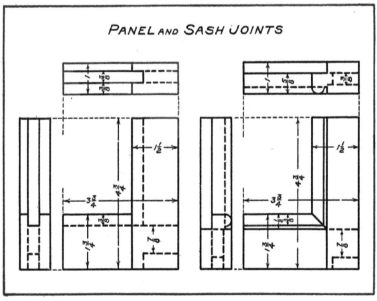

PANEL AND SASH JOINTS

PLATE 29

DETAILS OF PANEL AND SASH JOINTS

219. Example.

This plate shows three views of each rail and each stile of the joints shown on Plate 28.

Purpose.—The object of this plate is to show a suitable arrangement of the details and views of objects similar to those shown on Plate 28. The problems given in connection with this plate, will give drill and practice in analyzing projects which must of necessity have hidden lines to make possible a clear understanding of their construction.

Problem 1.—Make a drawing of the front and top views as shown in the example, and show a revolved section in each front view.   *Time, 4-1/2 hours.*

Problem 2.—Make a drawing showing the front, bottom, and left-hand end views, of the details shown in the example.
                                 Time, 4-1/2 hours.

Problem 3.—Make a detail drawing showing two views of each part of Problem 3, Plate 28.   *Time, 5 hours.*

PLATE 29.

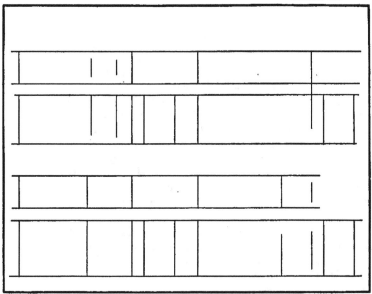

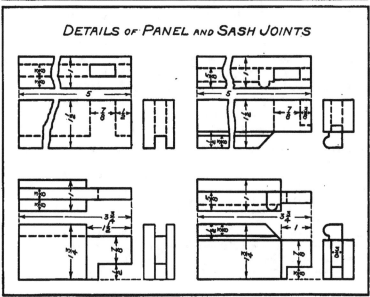

DETAILS OF PANEL AND SASH JOINTS

PLATE 30

MACHINE PARTS

220. Example.

This plate shows three views of each, of two cast-iron or steel machine parts. The objects are placed in a definite relation to the three planes of projection. The one at the left is 1/2 inch from V, 1/2 inch from P, and has its upper surface 5/16 inch from H. The object at the right is placed 5/8 inch from V, in contact with P, and has its upper surface 1/2 inch from H.

Purpose.—The purpose of this plate is to show cylindrical objects having flat surfaces and rectangular and round holes. The plate furthermore illustrates some fine points in projection, showing how a front view and end view may be drawn from the top view.

Problem 1.—Make a drawing showing three views of each object in the example, but turned through an angle of 90° about their vertical axes. Show both objects 3/4 inch from V, 1/2 inch from H, and 7/8 inch from P.   *Time, 5 hours.*

Problem 2.—Invert the front views of both objects shown in the example, and project the top and end views. Let both objects be removed 1/2 inch from V, 3/4 inch from H, and in contact with P.   *Time, 5-1/2 hours.*

Problem 3.—Make a drawing showing three views of each object, shown in the example, turned through an angle of 45° in a clockwise direction about their vertical axes. Let the centers of both objects be 1-5/8 inches from V and 1-3/4 inches from P; also let the objects rest on an auxiliary horizontal plane which is 4-1/2 inches from H.   *Time, 5-1/2 hours.*

Optional Practice.—These problems afford an opportunity for shade lines, also for line shading.

Optional Practice.—Calculate the volumes, and weights of the objects drawn, if made of cast iron.

NOTE.—To find the volume of the object on the left see Arts. 337 and 342. The volume of the rectangular hole may be taken as being equal to 9/16 × 1-1/2 × 1-1/2.

To find the volume of the object on the right see Arts. 329 and 342. The length of the chord and the height of the segment, for the cut-out portions, must be measured on the drawing.

The volume multiplied by .26 will give the weight.

PLATE 30.

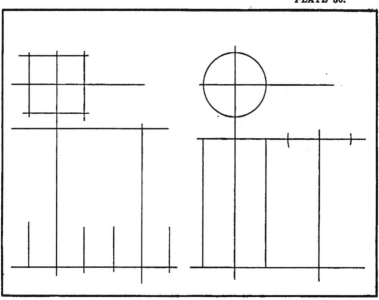

MACHINE PARTS

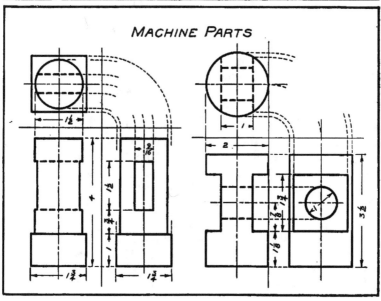

PLATE 31

HANDWHEEL

221. Example.

This plate shows two views of a handwheel, such as is used for opening or closing a valve, or for adjusting the spindle of a lathe. Small handwheels may be made of either cast iron or of brass, while large handwheels are generally made of cast iron.

Purpose.—The object of this drawing is to give an example in conventional cross-sectioning. By referring to the end view, it will be seen that the arm of the handwheel is not sectioned, although the section is taken on a vertical plane passing through the axis. The problems in connection with this plate will give practice in finding the centers of a number of circular arcs, which, in order to present a good appearance, must be exactly tangent to the straight lines of the arms. Problems 2 and 3 will give practice in laying out an odd number of arms in wheels or in pulleys.

Problem 1.—Make a drawing showing two views of a 5-inch handwheel having a rim 11/16 inch in diameter. All other dimensions are to be as shown in the example. Place the front view with the center lines of the arms making angles of 45° with a horizontal line, and project the end view. Show the end view in full cross-section.   *Time, 5–1/2 hours.*

Problem 2.—Design and draw two views of a handwheel having three arms, and of the following dimensions:

Outside diameter, 7″.	Hub projecting beyond the rim, 1/2″.
Diameter of rim, 3/4″.	Length of hub, 1–1/4″.
Diameter of shaft, 7/8″.	Width of arm at the rim, 5/8″.
Diameter of hub, 1–5/8″.	Width of arm at the hub, 7/8″.

Draw radii of circular arcs as in the example. Show the end view in half cross-section.   *Time, 5–3/4 hours.*

Problem 3.—Design and draw two views of a handwheel having five arms, and of the following dimensions:

Outside diameter, 8″.	Hub projecting beyond the rim, 5/8″.
Diameter of rim, 3/16″.	Length of hub, 1–3/8″.
Diameter of shaft, 1″.	Width of arm at the rim, 11/16″.
Diameter of hub, 1–3/4″.	Width of arm at the hub, 1″.

Draw radii of circular arcs as in the example. Show the end view in half cross-section, and also show a revolved section on one arm in the front view.   *Time, 6–1/2 hours.*

PLATE 31.

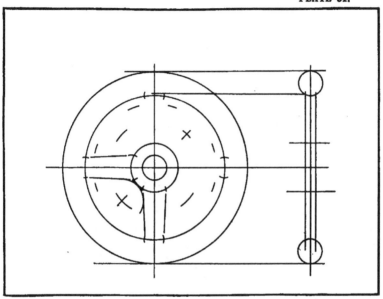

HAND WHEEL

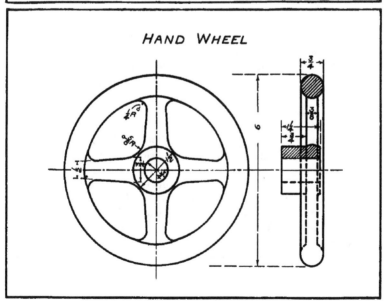

PLATE 32

STEP LADDER

222. Example.

This drawing shows a front view, a side view, a part top view, and a detail, of a step ladder. Ladders similar to the one shown are generally made of poplar wood, although any suitable wood may be used.

Purpose.—The purpose of this plate is to show an object drawn to a reduced scale. The tie bar shown at the right is drawn to a scale different from that used for the ladder; this practice is permissible and is often resorted to in practical work. Since both sides of the ladder are alike, it is necessary to show only a part of the top view, which will give all the information needed. The drawing also shows that, when an object is made of a uniform thickness of stock, the dimensions need not necessarily be placed on each piece, but may be specified in a note.

The solution of the problems in connection with this plate will give an opportunity for a limited amount of design.

Problem 1.—Make a drawing to a scale of 1-1/2 inches equals 1 foot of the three views, and 3 inches equals 1 foot of the detail, of the ladder shown in the example.   *Time, 7 hours.*

Problem 2.—Make a drawing showing the side and front views and detail of a ladder as shown in the example, but having an additional *tread*. Total height to be 3 feet 9 inches, and the tie bar to be increased 4 inches in length. All other dimensions are to be as shown in the example. Draw to a scale of 2 inches equals 1 foot.   *Time, 7-1/2 hours.*

Problem 3.—Design and draw the necessary views of a ladder, similar to the example, having six treads, each of 9 inch rise. Let the extreme width of the ladder be 20 inches. Make the *stringers* 5 inches wide, and back supports and ties 2 inches wide, and let the thickness of all stock be 1-1/8 inches. Draw to a scale 1-1/2 inches equals 1 foot.   *Time, 8 hours.*

Optional Practice.—Make a bill of material showing the sizes of the pieces; also find the number of square feet of stock required, for the object drawn.

PLATE 32.

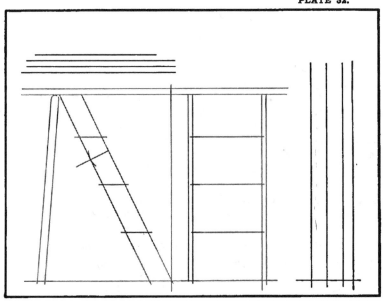

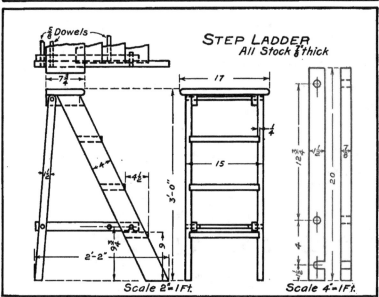

STEP LADDER
All Stock ⅞" thick

Scale 2"=1Ft.   Scale 4"=1Ft.

PLATE 33

SHAFT BEARING

223. Example.

This drawing shows three views of a bearing such as is used for supporting shafting, or a journal for machinery. Bearings of this kind are always made of cast iron, and are lined with an *anti-friction* metal, usually, babbitt metal.

Purpose.—The object of this plate is to familiarize the student with elementary machine parts, and also to give drill in reading hidden lines. Problems 2 and 3 will give practice in drawing similar bearings of different dimensions, which will give an opportunity for a limited amount of design.

Problem 1.—Make a drawing of the three views of the bearing shown in the example, and show the left-hand half of each view in section.   *Time, 7 hours.*

Problem 2.—Draw three views of a bearing similar to the one shown in the example, increasing the 1-1/2, 1-3/4, and 2-1/2 inch diameters to 2, 2-1/4, and 3 inches respectively; all other dimensions to be as shown. Show the front and end views in half cross-section.   *Time, 7 hours.*

Problem 3.—Design and draw three views of a bearing for a 2 inch shaft. Find the dimensions by proportion, using the diameters as constant numbers.

Thus, to find the length of the base

$$2 : 1\text{-}1/2 :: X : 7 \quad \therefore X = \frac{2 \times 7}{1.5} = 9.33 \text{ inches.}$$

To find the height to the center

$$2 : 1\text{-}1/2 :: X : 1\text{-}7/8 \quad \therefore X = \frac{2 \times 1\text{-}7/8}{1\text{-}1/2} = 2.5 \text{ inches.}$$

All other dimensions may be found similarly, and may be changed to the nearest 1/16 inch, should they occur in decimals.

Show the front and end views in half cross-section.

Time, 7-1/2 hours.

Optional Practice.—Find the approximate weight of the bearing drawn, if made of cast iron.

NOTE.—The weight may be found very closely by first finding the volume of the base, neglecting the rounded corners, and subtracting the volume of the two bolt holes, plus the volume of the hollow cylinder forming the bearing, neglecting the two "lips" for holding the babbitt metal. Multiplying the volume by .26 will give the weight.

PLATE 33.

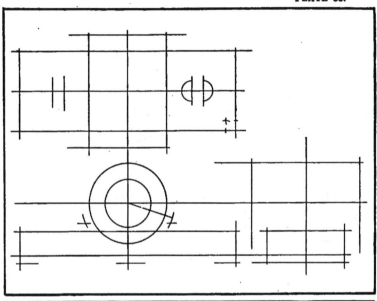

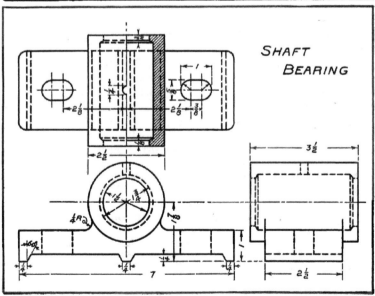

SHAFT
BEARING

PLATE 34

HEXAGONAL TABOURET

224. Example.

This plate shows two views of a hexagonal tabouret with a revolved section of one side showing the construction of the joints. Tabourets are used for supporting flower pots or jardinieres; they are usually constructed of quartered oak.

Purpose.—The object of this plate is to show how an object may be represented by a front and an end view and without a top view. The title of the drawing gives the shape of the top view, while the revolved section shows the thickness of the stock from which the body is to be made. The thickness of the body may also be given in a note, in that case eliminating the revolved section. The layout furthermore illustrates a method for projecting semicircular or other curves on oblique surfaces. It also shows how to obtain the front and side views of a hexagon by the aid of a few preliminary pencil lines. Practice will be afforded in drawing curves, with the aid of irregular curves.

Problem 1.—Design and draw the front and end views, also show a revolved section, of a hexagonal tabouret having the following dimensions:

Height, 14 inches; width of sides, 5 inches; width of arches, 3-1/2 inches. Total height of arches, 9 inches. Thickness of stock of the sides, 3/4 inch; thickness of top, 7/8 inch. Let the top project beyond the sides 1 inch. Draw to a scale of 6 inches equals 1 foot. *Time, 7-1/2 hours.*

Problem 2.—Draw the front and bottom views of a tabouret having the dimensions as given in Problem 1. *Time, 8 hours*

NOTE.—The vertical axis of the tabouret must be placed parallel to the horizontal border line of the drawing. Place the front view on the right side and the bottom view on the left side of the drawing.

Problem 3.—Design and draw any two views of a tabouret having five sides, and of the dimensions as given in Problem 1.

Time, 10 hours.

NOTE.—For the construction of a regular pentagon see GEOMETRICAL PROBLEMS (Art. 261).

Optional Practice.—Make a bill of material showing the dimensions and number of pieces required for the problem drawn. Also find the number of square feet of stock required.

PLATE 34.

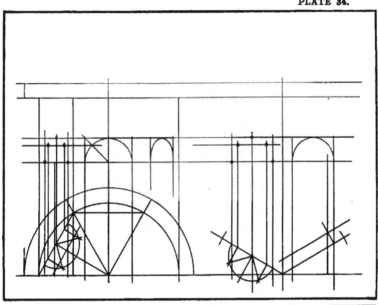

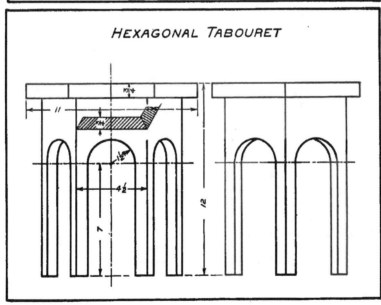

HEXAGONAL TABOURET

PLATE 35

CRANE-HOOK

225. Example.

This drawing shows the front and end views, also three cross-sections, of a steel drop-forged crane-hook.

Purpose.—This plate affords practice in careful analysis of a working drawing. It will be seen when studying the example, that many points must be taken into consideration to achieve good results. The problems given in connection with this plate will afford practice in laying out angles with the protractor, and in drawing smooth curves consisting of a number of circular arcs, whose centers must be accurately found. Problems 2 and 3 will give practice for obtaining the dimensions of similar hooks by proportion.

Problem 1.—Make a drawing showing the front view of the crane-hook shown in the example, revolved about its vertical axis through an angle of 180°, so that the throat opening is on the right, and project a right-hand view. Also show three cross-sections.   *Time, 8–1/4 hours.*

Problem 2.—Design and draw two views, similar to the example, for a crane-hook having a 1–1/2 inch throat opening. Find the dimensions by proportion and show them to the nearest thirty-second of an inch.

Thus, for the total length l

$$1\text{--}1/2 : 1\text{--}1/4 :: l : 7 \quad \therefore l = \frac{1\text{--}1/2 \times 7}{1\text{--}1/4} = 8.4 = 8\text{--}13/32 \text{ nearly.}$$

Other dimensions may be found similarly, or if we divide 1–1/2 by 1–1/4 we get 1.2 which may be used as a multiplying constant for all the dimensions given in the example.

For the total length of the hook, l.

$l = 1.2 \times 7 = 8.4$, as before.

For the largest radius, r.

$r = 1.2 \times 3\text{--}3/8 = 4.05 = 4\text{--}1/16$, nearly.

Distance between the two centers in the throat, d.

$d = 1.2 \times 1/4 = .3 = 5/16$, nearly.   *Time, 9 hours.*

NOTE.—The length of a number of the radii, obtained by calculation, may have to be slightly modified as the drawing proceeds.

Problem 3.—Study Problem 2. Find the multiplying constant for a crane-hook having a 2–1/4 inch throat opening, and draw front and end views; also show three cross-sections. Use a scale of 6 inches equals 1 foot and express all dimensions in inches and decimals of an inch, correct to two places. *Time, 9 hours.*

PLATE 35.

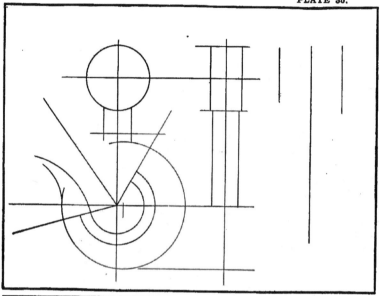

CRANE-HOOK

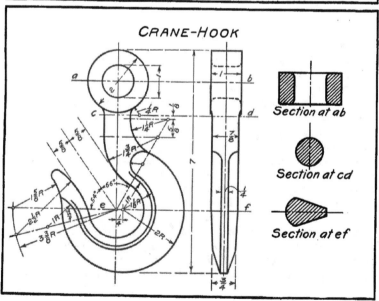

Section at ab

Section at cd

Section at ef

PLATE 36

PRINTING FRAME

226. Example.

This plate shows two views of a printing frame, such as is used for making blue prints or for printing from photographic negatives. The drawing shows the glass, pressure cover, and springs in place.

Purpose.—The object of this plate is to show an assembly drawing of an object consisting of a number of different parts and different materials. The plate also shows a number of conventional cross-sections showing the shape of the frame, the glass, and the method of fastening the end cleats to the cover. A portion of the cover is broken away to show the "lip" on the frame in full lines.

NOTE.—This plate will afford an opportunity for reading and analyzing a working drawing. The dimensions of the hinges are not given, as any standard hinge of suitable size may be used.

Problem 1.—Make *two* drawings of the printing frame shown in the example. The first is to contain two views of the frame, having the cover and glass removed, and showing two revolved sections in the top view. The second drawing is to show two views of the cover, springs, and hinges, with a revolved section showing the "tongue-and-groove" joint. Scale, 6 inches equals 1 foot.   *Time, 9-1/2 hours.*

Problem 2.—Make a drawing showing two views of a 5 × 7 inch printing frame similar to the example. Use the number .6 for the multiplying constant to obtain the dimensions for the cross-section of the frame, and width of springs. The pressure cover is to be 3/8 inch thick and consist of two parts, one being 3 inches long and the other 4 inches long. Details are left to the student's judgment.   *Time, 10 hours.*

NOTE.—A 5 × 7 inch printing frame should measure about 5-1/8 × 7-1/8 inches, so that the glass plate will not bind in the frame.

Problem 3.—Design and make a drawing for a printing frame to make 15 × 22 inch blue prints. For the dimensions of the cross-section of the frame, and width of springs, use 1.3 as a multiplying constant. Make the pressure cover 5/8 inch thick and show it cut into three equal parts; this will require three springs. The details of design are left to the student's judgment. Draw to a scale of 3 or 4 inches equals 1 foot.
                    Time, 10-1/2 hours.

PLATE 36.

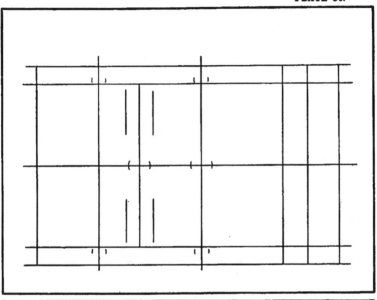

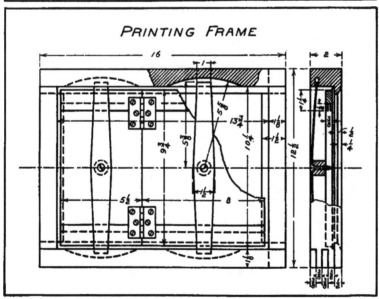

PRINTING FRAME

PLATE 37

PULLEY

227. Example.

This plate shows two views of a 9 inch cast-iron pulley, with five curved arms. Pulleys are used for the transmission of power by means of leather or other belting.

Purpose.—The object of this plate is to show how circular objects like pulleys, gears, cylinders, etc., may be drawn showing a portion broken away. This practice is frequently resorted to by draftsmen, when the standard size sheet of paper is not large enough to permit making the full drawing.

Problem 1.—Draw two views of a pulley of the dimensions as shown in the example, but having four arms instead of five. Show the end view in half section, and also show a revolved elliptic cross-section on one arm in the front view.

Time, 9 hours.

Problem 2.—Draw two views of a pulley having five straight arms. Let the arms be 1 inch wide at the center of the hub and 3/4 inch wide at the inside of the rim, also let the hub be 1–3/4 inches in diameter and the bore for the shaft be 7/8 inch; all other dimensions to be as shown in the example. Show the end view in half cross-section. *Time, 10–1/2 hours.*

Problem 3.—Design and draw two views of a 10 inch pulley having five curved arms. Show the end view in half cross-section. Find the dimension by proportion from those given in the example. *Time, 12 hours.*

Thus, for the width of face, x.

$$x : 2 :: 10 : 9 \quad \therefore \quad x = \frac{20}{9} = 2.22$$

For radius of arm, y.

$$y : 2.312 :: 10 : 9 \quad \therefore \quad y = \frac{23.12}{9} = 2.56$$

Other dimensions are found similarly.

Optional Practice.—Find the circumference of the rim of the pulley drawn. (See Art. 322.)

Find the diameter of a pulley whose circumference is 34.557 inches. (See Art. 325.)

Find the diameter, in inches, of a pulley whose circumference is 4.712 feet. (See Art. 325.)

Find the surface speed, in feet, of a pulley 12 inches in diameter, making 200 revolutions per minute.

NOTE.—Multiply the circumference of the pulley in feet, by the revolutions per minute.

PLATE 37.

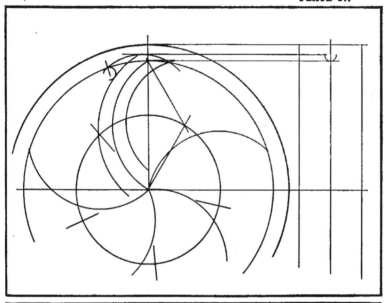

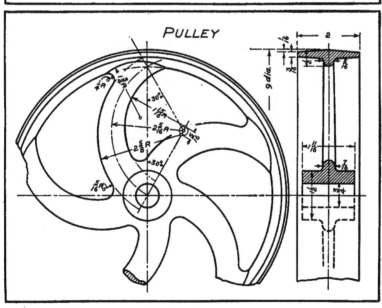

PULLEY

PLATE 38

BOLTS AND NUTS

228. Example.

This plate shows a number of views of two standard size bolts and nuts, such as are used in machine construction.

Purpose.—The object of this plate is to show how a number of views of an object may be projected. The problems in connection with this plate will give the student practice in the use of equations or formulas, such as are found in many books on machine design.

The proportions for the various dimensions of bolts and nuts can be found from the appended table.

Let D = the diameter of the bolt.
Then $A = 1.5D + .125.$
      $B = 1.75D + .125.$
      $C = .75D + .0625.$
      $E = .707A.$
      P = Pitch of thread.
      N = Number of threads per inch = $1/P.$
      $R = .875D.$
      $R' = 1.125D.$
      S is to be found by trial.

Problem 1.—Make a drawing showing the views of two bolts as in the example.   Let D = 7/8 inch and N = 9.   *Time,* 10 *hours.*

Problem 2.—Make a drawing showing the views of two bolts as shown in the example.   Let D = 1-1/4 inches and N = 7.

Time, 10–1/2 *hours.*

Problem 3.—Make a drawing showing the views of two bolts as shown in the example.   Let D = 1-3/8 inches and N = 6.

Time, 11 *hours.*

Note 1.—The drawings in the example are for bolts 1 inch in diameter and 4 inches long, having 8 threads per inch.

Note 2.—The length of a bolt is measured from the under side of the head to the end, and is dependent upon the use for which the bolt is intended.

PLATE 38.

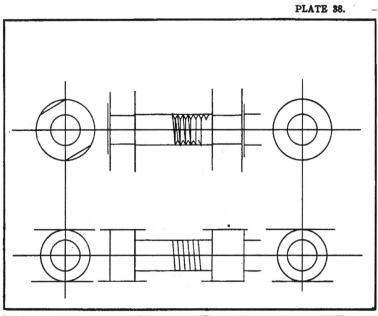

BOLTS AND NUTS

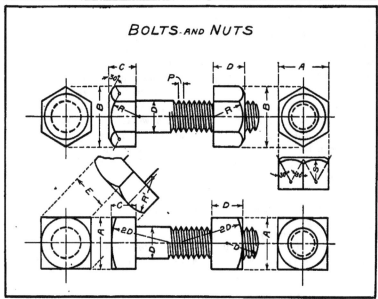

CHAPTER VIII

GEOMETRICAL PROBLEMS

229. Introductory.

Geometrical drawing as pointed out in the introduction, is of great value to the mechanical draftsman in practice.

The problems here presented are a series of selected exercises. A working knowledge of many of these is indispensable to the good draftsman; the whole series covers practically the field of geometrical drawing.

The problems should be treated only as supplementary to our main purpose, which is mechanical drawing. A judicious selection of problems must be made by the student or the instructor, should he wish to include some geometrical drawing in his course. A brief series of from twelve to eighteen of these problems incorporated in a regular course of mechanical drawing will ensure a working knowledge of geometrical drawing. The selection from the large number of problems offered should be made to suit the individual requirements.

In working these problems it is of the highest importance that the drawings should be neat and accurate. They need not necessarily be inked, if time is of importance, as lines of various strengths may be drawn with the pencil. If finished in pencil, or in black ink only, the given lines should be of medium weight, the required lines heavy, and all construction lines should consist of short fine dashes. Geometrical drawings, however, are most easily read when they are made by drawing the given lines with blue ink, all construction lines with red ink, and the required lines with black ink. When colored inks are used, all lines may be drawn continuous and of uniform width.

SIMPLE GEOMETRICAL PROBLEMS

230. To Bisect a Given Line. (Fig. 86.)

With A as a center and a radius R, greater than half of AB, draw an arc abc. With the same radius and B as a center, draw

176

an arc intersecting the first arc in the points *a* and *c*. Draw a line from *a* to *c*. This line will be the bisector of line *AB*.

231. To Bisect a Given Arc. (Fig. 87.)

With *A* and *B* as centers and a radius *R*, greater than half the length of the given arc, draw the small arcs *a* and *c*, intersecting at *e*. With *A* and *B* as centers and the same radius, or of radius *R'*,

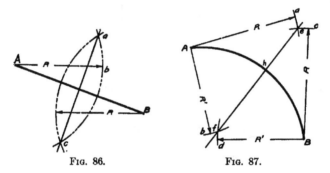

FIG. 86.   FIG. 87.

draw small arcs *b* and *d*, intersecting at *f*. A line drawn through the intersections *e* and *f* will bisect the arc at *h*.

This is essentially the same as bisecting a straight line. The small arcs *a* and *c* may be extended as continuous arcs to *b* and *d*, but this is not necessary.

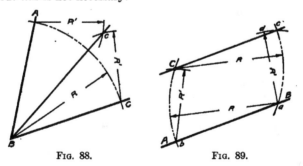

FIG. 88.   FIG. 89.

232. To Bisect a Given Angle. (Fig. 88.)

Let *ABC* be the given angle. With *B* as a center draw an arc of radius *R* cutting the angle at the points *A* and *C*. With *A* and *C* as centers and *R'* as radius draw small arcs, intersecting at *c*. From *c* draw a line to the vertex *B*. This line is the bisector of the angle.

233. To Draw a Line Parallel to a Given Line through a Given Point. (Fig. 89.)

Let AB be the given line and C the given point. With C as a center and R as a radius draw an arc cutting the given line at a. From a, and with the same radius, draw an arc cutting AB at b; with bC as a radius and a as a center draw the small arc d cutting the arc ac. Through the intersection of these two arcs and the point C draw a line. This line will be parallel to the given line.

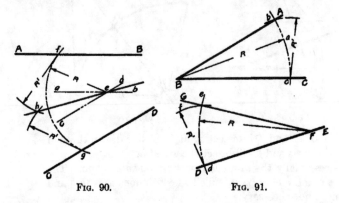

FIG. 90.   FIG. 91.

234. To Divide an Angle, whose Sides do not Meet, into Two Equal Parts. (Fig. 90.)

Let AB and CD be the given angle. Draw lines ab and cd parallel to and at equal distances from the given lines. From their point of intersection e draw an arc cutting the given lines at f and g. From these points as centers and with radius R' draw small arcs intersecting at h. A line drawn through h and e will divide the given angle into two equal parts.

235. To Construct an Angle Equal to a Given Angle, on a Given Line. (Fig. 91.)

Let ABC be the given angle and DE the given line. With any convenient radius as R, draw an arc cutting the angle at b and c. From the point F, on the given line, and with the same radius draw the arc de. From d as a center, with radius cb, draw the small arc f cutting the arc de. From F, and through this point of intersection, draw the line FG.

236. To Draw a Line Perpendicular to a Given Line from a Point on the Line, or Outside the Line. (Fig. 92.)

Let C be a point outside the given line AB. From C at a center and radius R, draw an arc cutting the given line in a and b. With a and b as centers and radius R'' draw small arcs c and d. Through their intersection and the point C draw the line EF. This line will be perpendicular to the given line.

Let the point, as F, be on the given line. With F as a center draw small arcs, of the same radius, cutting the given line at a and b. With these points as centers and a little larger radius, draw small arcs g and h. Through the intersection of g and h, draw the line EF, which will be the perpendicular required.

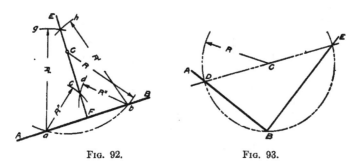

FIG. 92.   FIG. 93.

237. To Draw a Perpendicular at the End of a Given Line. (Fig. 93.)

Let AB be the given line. Take any point without the line as C, and with CB as a radius draw an arc somewhat longer than a semicircle, intersecting AB at D. Through the points D and C draw a line intersecting the arc at E. Then a line drawn from B through E will be the perpendicular required.

Since line DCE is a diameter of arc DBE, angle ABE is a right angle.

238. To Divide a Given Line into any Number of Equal Parts. (Fig. 94.)

If AB is the given line, draw any line from A making any convenient angle with AB. From the point A mark off the required number of equal parts, of any convenient length, as Ab, bc, cd, etc. Draw a line from the last point as g, to B. Then draw lines parallel to gB from all the points on the auxiliary line, cutting the given line at points f', e', d', etc. This will divide the given line into the required number of equal parts.

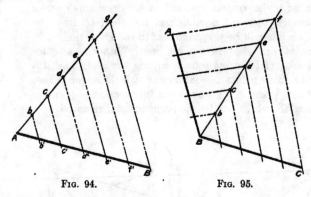

Fig. 94.   Fig. 95.

239. To Divide the Sides of an Angle into the Same Number of Equal Parts. (Fig. 95.)

Let ABC be the given angle. From B draw any convenient line as Bf. Mark off on this line the number of equal divisions required. Proceed as in the preceding problem. This will divide the sides of the angle ABC into the required parts.

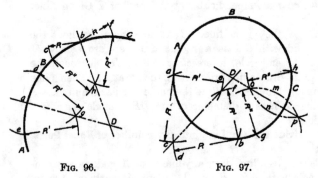

Fig. 96.   Fig. 97.

240. To Locate the Center of a Given Arc. (Fig. 96.)

Let ABC be the given arc. From any two points on the arc as a and b, and with any convenient radius as R, draw small arcs *e, d, c*, and *f*, intersecting the given arc. From these points of intersection and with a slightly larger radius, draw arcs intersecting at *g* and *h*. The intersection *D*, of lines drawn through *ag* and *bh*, will be the required center.

The problem can also be solved by drawing any two chords to the given arc, bisecting these chords, and erecting perpendiculars whose intersection will be the center desired.

241. To Find the Center of a Given Circle. (Fig. 97.)

Let *ABC* be the given circle. From any two points, as *a* and *b*, draw intersecting arcs *c* and *d*, also *e* and *f*. From two other points *g* and *h* on the circle, draw arcs *m* and *n* intersecting at *o* and *p*. The intersection *D*, of the two lines drawn through the intersecting arcs, will be the required center.

This problem can be solved as the foregoing problem. Instead of drawing small arcs from points *a* and *b*, continuous arcs may be drawn.

LINES AND TANGENT ARCS

242. In mechanical drawing it is very often necessary to draw curves, or circular arcs, tangent to straight lines or curves. While the methods for making such constructions are purely geometrical, the processes employed are not generally dealt with in geometries. The experienced draftsman can, and very often does, find the center of an arc by trial, and then estimates the points of tangency of a curve and line. Not wishing to take the time to make a geometrical construction, the draftsman very often loses time in making assumptions. Most of the geometrical constructions which are used in technical drawing are very simple and should be studied and mastered by the student so that he can apply them when needed without finding it necessary to refer to the book. The beginner will very early in the course discover for himself that the neatness of a drawing will depend, in a large measure, upon obtaining good tangencies between lines and curves.

243. To obtain good results it will be necessary to remember the following rules:

First.—In joining a curve to a straight line, the center of the curve must lie on a line which is perpendicular to the given line with which the curve is to form a tangent. (Figs. 98 and 99.)

Second.—In joining a straight line to a curve, the straight line must be perpendicular to a radial line drawn to the point of tangency. (Figs. 98 and 99.)

Third.—The point of tangency of two curves will lie on a line joining the centers of the curves. (Figs. 101, 103, 104 and 105.)

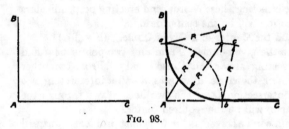

Fig. 98.

244. To Draw an Arc of Given Radius, Tangent to Two Given Lines which are at Right Angles. (Fig. 98.)

Let AB and AC be the given lines. From A as a center draw an arc, having the required radius, intersecting the lines at a and b. With a and b as centers and the same radius, draw intersecting small arcs c and d. The point of intersection will,

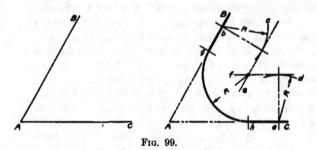

Fig. 99.

be the center of the arc, of radius R, which is to be tangent to the two lines.

NOTE.—The point of tangency will be at points a and b, and the intersection of the two small arcs is on lines drawn perpendicular to the given lines from the same point.

245. To Draw an Arc of Given Radius Tangent to Two Lines which are not at Right Angles. (Fig. 99.)

Let AB and AC be the given lines. With the given radius R

and from any convenient points as a and b, on the given lines, draw small arcs c and d. The intersection of lines e and f, drawn tangent to the arcs and parallel to the given lines, will be the center of the required arc.

NOTE.—The point of tangency of the lines and arc will be at points g and h, which lie on perpendiculars drawn from the center of the arc to the given lines.

246. To Draw a Circular Arc of Given Radius Tangent to a Given Circular Arc and Given Straight Line. (Fig. 100.)

Let AB be the given arc and CD be the given line. From the center of the given arc, draw an arc ab whose radius is equal to the radius of the given arc plus the radius of the required arc. Also draw a line cd parallel to the given line at a distance equal

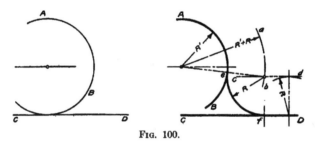

FIG. 100.

to the radius of the required arc, by the method used in Fig. 99. The intersection of the arc ab and the line cd will be the center of the required arc.

NOTE.—The point of tangency of the two arcs is on a line joining the centers of the two arcs. Also the point of tangency of the arc and straight line is at the foot of a perpendicular from the found center to the line. ·

247. To Draw a Circular Arc, Tangent at a Given Point on a Circular Arc, also to a Straight Line. (Fig. 101.)

Let ABC be the given arc and B the given point; also let DE be the given line. At the point B draw a line fg perpendicular to a line hj, which passes from the center of the given arc through the point B, until it intersects the given line at k. Bisect the angle fkE. The point of intersection of line hj and the bisector of the angle is the center of the arc to be found. The points of tangency are at B and m.

248. To Draw a Circular Arc Tangent to Three Given Lines. (Fig. 102.)

Let *AB*, *BC*, and *CD* be the given lines. Bisect angle *BCD*, by the usual method. Angle *ABC* being a right angle, a 45°

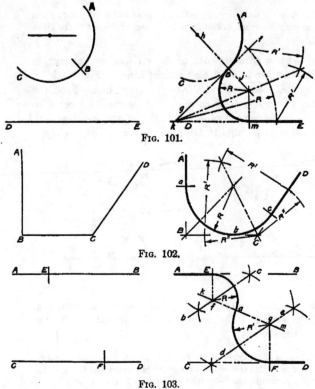

FIG. 101.

FIG. 102.

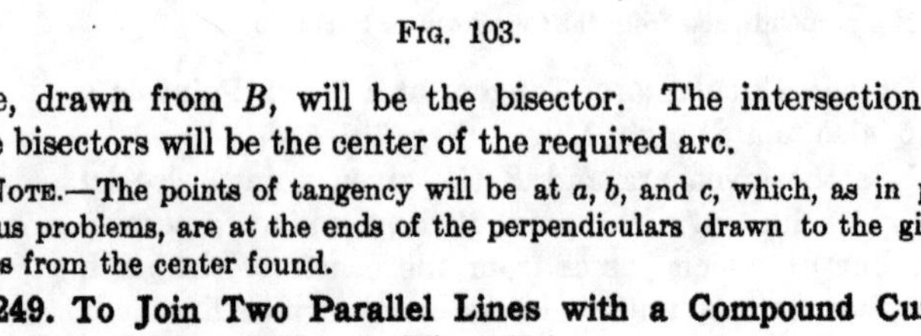

FIG. 103.

line, drawn from *B*, will be the bisector. The intersection of the bisectors will be the center of the required arc.

Note.—The points of tangency will be at *a*, *b*, and *c*, which, as in previous problems, are at the ends of the perpendiculars drawn to the given lines from the center found.

249. To Join Two Parallel Lines with a Compound Curve Tangent at Given Points. (Fig. 103.)

Let *AB* and *CD* be the given lines, and *E* and *F* be the given points. Join *EF* by a straight line. Select any point as *a*.

Draw perpendiculars, from points E and F, to the given lines. Bisect the distances Ea and aF, and draw perpendiculars bc and de. The intersection of the two sets of perpendiculars will give the centers f and g for the required curve.

NOTE.—The points of tangency of the curves and lines are at E and F. The point of tangency of the curves is at a on the line km joining the centers of the curves.

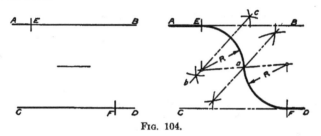

FIG. 104.

250. To Join Two Parallel Lines with a Smooth Curve Tangent at Given Points. (Fig. 104.)

Let AB and CD be the given lines, E and F the given points. Bisect a line drawn from E to F. Also bisect Ea and draw perpendicular bc. Draw a perpendicular from E intersecting line bc; the point of intersection will be the center of the upper curve.

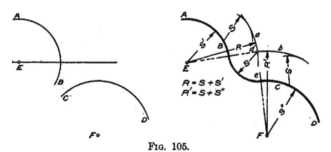

FIG. 105.

The center of the lower curve may be found in the same way, or a line drawn from the center found, through the point a, intersecting a perpendicular from the point F, will give the second center.

NOTE.—The point of tangency of the two arcs is on a line joining their centers.

251. To Draw an Arc of Given Radius Tangent to Two Given Circular Arcs. (Fig. 105.)

Let AB and CD be the given arcs; also E and F their centers. With radius R, equal to the radius of the arc AB plus the radius of the required arc, draw small arc a. Draw arc b with radius R' equal to the radius of arc CD plus the radius of the required arc. The intersection d, of arcs a and b, will be the center of the required arc.

NOTE.—The points of tangency will be on lines Ed and eF which connect the centers of the arcs.

REGULAR POLYGONS

252. Of the polygons given, the triangle, square and hexagon are those most commonly used. The hexagon is nearly always used when drawing the head of a hexagonal bolt or nut.

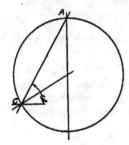 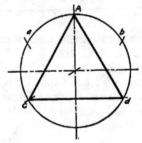

FIG. 106.

253. To Inscribe a Regular Triangle in a Given Circle. (Fig. 106.)

Geometrical Method.—From A, draw arcs a and b with a radius equal to the radius of the circle, and from points a and b, draw arcs c and d with the same radius. Join Ac, Ad and cd which will give the required triangle.

NOTE.—This triangle is equilateral, also equiangular.

Draftsman's Method.—From point A on the circle draw 60° lines with the triangle, intersecting the circle at c and d. With the T-square draw line cd thus completing the triangle.

254. To Inscribe a Square in a Given Circle. (Fig. 107.)

Geometrical Method.—Draw two diameters at right angles to each other, intersecting the circle at A, B, C and D. Connecting these points will give the required square.

Draftsman's Method.—Draw two diameters through the center of the circle, at right angles, with the 45° triangle. Connect the points of intersection of the diameters and circle.

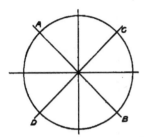 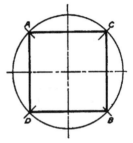

FIG. 107.

255. To Inscribe a Regular Pentagon in a Given Circle. (Fig. 108.)

Geometrical Method.—Draw two diameters AB and CD at right angles to each other. Bisect the radius eB. With f as a center and radius fC draw the arc Cgh, intersecting the diameter AB at h. With C as a center and Ch as a radius draw the arc IJK, intersecting the circle at I and K. The chords drawn from C, to I and K, are two sides of the pentagon. Find the remaining two

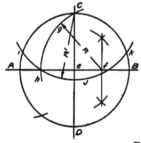 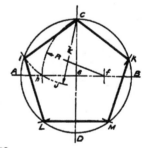

FIG. 108.

points by drawing small arcs from I and K, with radius R', intersecting the circle at L and M. Connect CK, ML, etc., which will give the required pentagon.

Draftsman's Method.—With the dividers estimate the length of a chord which will step five times around the circumference; adjust the dividers and stepping off until the proper divisions are obtained, connect the points thus found.

256. To Inscribe a Regular Hexagon in a Given Circle.
(Fig. 109.)

Geometrical Method.—From *A* and *B*, and with a radius equal
to that of the circle, draw small arcs intersecting the circle at

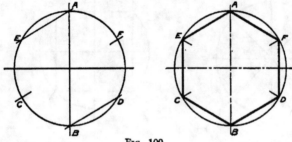

FIG. 109.

C, D, E and *F*. Connect the points by chords which will give
the required hexagon.

Draftsman's Method.—Draw *AB* with the triangle. With
the 30° triangle draw *EA* and *BD*. Reverse the triangle and draw
BC and *AF*. Draw vertical lines with triangle from *C* to *E* and
D to *F*, completing the figure.

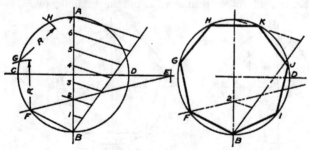

FIG. 110.

257. To Inscribe a Regular Heptagon in a Given Circle. (Fig.
110.)

Geometrical Method.—Draw two diameters at right angles to
each other. Divide *AB* into seven equal parts, by the method
used for dividing a line into any number of equal parts. Extend
the diameter *CD* to *E*, making *DE* equal to three-quarters of the
radius of the given circle. Draw a line from *E* through point

2 intersecting the circle at F. The chord BF will be one side of the required figure. With BF as a radius and F as a center draw small arc G; with G as a center and the same radius draw small arc H. Points I, J and K are found similarly, and connecting these points will give the heptagon.

Draftsman's Method.—Same as above.

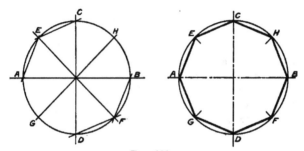

FIG. 111.

258. To Draw a Regular Octagon in a Given Circle. (Fig. 111.)

Geometrical Method.—Draw diameters AB and CD at right angles. Bisect the right angles thus formed and draw diameters EF and GH. Connecting AE, EC, CH, etc., by chords will give the figure required.

Draftsman's Method.—With the T-square draw diameter AB. With 90° triangle draw CD, and with 45° triangle draw EF and GH. Connect the points of intersection with the circle, by chords.

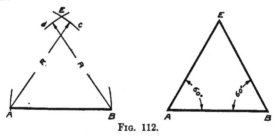

FIG. 112.

259. To Construct an Equilateral Triangle having Given the Length of One Side. (Fig. 112.)

Geometrical Method.—Let AB be the given side. From A as a center and radius AB, describe small arc c. With B as a center and the same radius, describe small arc *d*, intersecting arc *c* at *E*.   Join *EA* and *EB* which will give the required triangle.

Draftsman's Method.—With the 60° triangle draw *AE*; also by reversing the triangle draw *BE*.

NOTE.—Since the triangle is equilateral it is also equiangular.

260. To Construct a Square having Given the Length of One Side. (Fig. 113.)

Geometrical Method.—Let *AB* be the given length. At the point *B* draw a line perpendicular to *AB*. (See Fig. 93.) With *B* as a center and *BA* as a radius, draw small arc *C* intersecting the perpendicular drawn from *B*. With *A* and *C* as centers and the same radius, draw small arcs *d* and *e*, intersecting at *F*.   Join *AF* and *FC*.

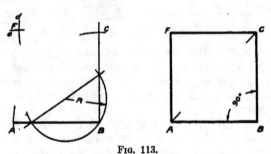

FIG. 113.

Draftsman's Method.—From *A* and *B* draw perpendiculars with the triangle. From *A* draw a 45° line cutting the perpendicular from *B* at the point *C*.   Through *C* draw a line parallel to *AB* with the T-square, giving the point *F*.

261. To Construct a Regular Pentagon upon a Line of Given Length. (Fig. 114.)

Geometrical Method.—Let *AB* be the given line. At *B* erect a perpendicular *BD*; with *B* as a center and *BA* as a radius, draw arc *AG* intersecting *BD* at *e*.   From *c*, the center of *AB*, and with *ce* as radius, draw an arc intersecting *AB* prolonged, at *f*.   With *A* as a center, and *Af* as a radius, draw an arc intersecting arc *Ae* at *G*.   Join *BG*, which will be one side of the pentagon.   Draw a line *GH* parallel to *AB*.   With *A* as a center and radius *AB* draw small arc *l*.   From the intersection of *GH* and *l*, also from the point *G*, with radius *R*, draw small arcs *j* and *k* intersecting at *L*.   Join the points found.

Draftsman's Method.—Same as above.

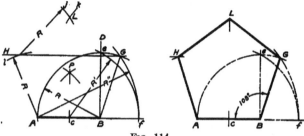

FIG. 114.

262. To Construct a Regular Hexagon upon a Line of Given Length.   (Fig. 115.)

Geometrical Method.—Let *AB* be the given length.   With *A* and *B* as centers and *AB* as radius, draw small arcs *c* and *d*. From points *A* and *B* draw lines *AF* and *BE*, passing through the intersection of arcs *c* and *d*.   Draw line *GH* parallel to *AB*.

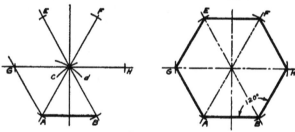

FIG. 115.

With *AB* as a radius and the intersection of arcs *c* and *d* as a center, draw small arcs, or a complete circle, intersecting the lines at points *G*, *E*, *F* and *H*, and connect the points.

Draftsman's Method.—From *A* and *B* draw *AF* and *BE* with the 60° triangle, and through their intersection draw a line parallel to *AB* with the T-square.   Complete the hexagon with the 60° triangle.

263. To Draw a Regular Heptagon on a Line of Given Length. (Fig. 116.)

Geometrical Method.—Draw a perpendicular *CD* at the center of *AB*.   With *A* as a center and *AB* as a radius, draw the arc *Ba* intersecting CD at b. Divide bA into six equal parts. With b as a center and $b1$ as a radius, draw an arc intersecting CD at G. With G as a center and GA as a radius, draw the circle ABD. Lay off with a small arc, the chordal distance AB, on the circle and connect the points thus found.

Draftsman's Method.—The same as geometrical method.

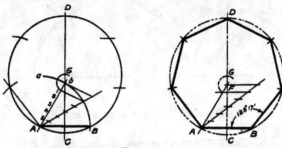

FIG. 116.

264. To Draw a Regular Octagon on a Given Base. (Fig. 117.)

Geometrical Method.—Let AB be the given base. At the center of AB draw the perpendicular CD. With C as a center and CA as a radius, draw an arc intersecting CD at a. With a as a center and radius aA, draw an arc intersecting CD a F.

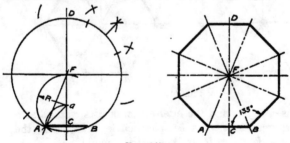

FIG. 117.

With F as a center and radius FA draw a circle. With AB as a chordal length, lay off small arcs on the circle and connect the points of intersection.

Draftsman's Method.—From A and B draw 45° lines equal in length to AB. At the extremities of these lines erect vertical lines, which will be two sides of a circumscribed square. Complete the figure with the 45° triangle.

265. To Inscribe a Polygon of Any Number of Sides in a Given Circle. (Fig. 118.)

The full geometrical method is given, which can be shortened by the draftsman according to his experience.

Let *ABCD* be the given circle. Through its center draw a horizontal line *DB*, also draw *AC* perpendicular to *DB*. With *C* as a center and *CA* as a radius draw arc *AF*, intersecting *DB* produced at *E*. Divide the diameter *AC* into as many equal parts as the polygon is to have sides. Draw a line from *E*

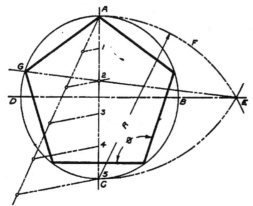

Fig. 118.—Regular polygon in a circle of given diameter.

through the point 2, intersecting the circle at *G*. The length of the chord *AG* will be the length of the sides of the polygon. This chordal distance repeated on the circumference of the circle will give all the points of the figure.

Note.—The line *GE* is drawn through the second division of the divided diameter for a polygon of any number of sides.

266. To Construct a Polygon of Any Number of Sides upon a Line of Given Length as One Side. (Fig. 119.)

Let *AB* be the given length. With *B* as a center and *BA* as a radius describe a semicircle, intersecting *AB* produced at *E*. With the aid of a protractor divide the semicircle into as many parts as the polygon is to have sides. Draw a line from *B* through the second division of the semicircle, intersecting it at 2. Bisect lines *AB* and *B2*, and draw perpendiculars *CK* and *GF*. With *H*, the intersection of the perpendiculars, as a center, and *HA* as a radius, draw the circumference of a circle. From *B* draw lines through the points on the semicircle, inter-

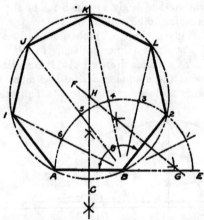

FIG. 119.—Regular polygon on a line of given length.

secting the circumference at points *I, J, K,* etc.   Join the points found on the circumference to complete the polygon.

NOTE.—The line from *B* is drawn through the second division from the right, for a polygon of any number of sides.

SPECIAL GEOMETRICAL CONSTRUCTIONS

267. To Draw a Smooth Curve Tangent to Two Given Points on the Sides of an Acute Angle.   (Fig. 120.)

Let *BAC* be the given angle, and *D* and *E* be the points to which the curve is to be tangent.   Divide *AD* into any number

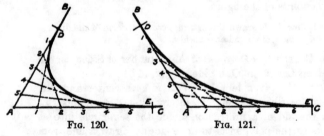

FIG. 120.   FIG. 121.

of equal parts, and divide *AE* into the same number of equal parts.   Number the points on *BA*, beginning with *D*, as 1, 2, 3, 4 and 5, also number *AC*, beginning with *A*, as 1, 2, 3,

4 and 5. From the numbered points on BA draw straight lines. to the corresponding numbers on AC. Draw a smooth curve from D, tangent to the intersecting lines, and ending at E, which will be the curve required.

268. To Draw a Smooth Curve Tangent to Two Given Points on the Sides of an Obtuse Angle. (Fig. 121.)

Let BAC be the given angle, D and E the given points. Divide the distances AD and AC into parts as described in the preceding problem. Draw a curve tangent to the lines.

ELLIPTIC ARCS

269. To Describe an Elliptic Arc with Three Centers, Having Given the Height, and the Length or Span. (Fig. 122.)

Let AB be the height and BC the span. Erect the perpendicular DE through the point F, the center of BC. On the line BC assume a point G, making GB somewhat less than the height EF. With BG as a radius and E as a center, draw a small arc I intersecting ED at J. Draw the line JG and at its center erect the perpendicular KL, intersecting ED at M. Arcs drawn with G as a center and GB as a radius, and M as a center and ME as a radius, will be tangent at H, on a line drawn through points M and G. The other side is constructed in a similar manner.

NOTE.—This curve is frequently used in architectural work.

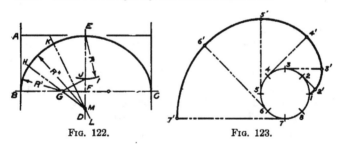

FIG. 122. FIG. 123.

INVOLUTE

270. An involute spiral is a curve described by the end of a string unwinding from a cylinder.

271. To Construct an Involute Spiral on a Given Circle as a Base. (Fig. 123.)

Divide the circumference of the given circle into any number of equal parts. Draw tangent lines at the points found. On the tangent drawn at 2, lay off the length of the chord 2 1, giving the point 2′. On the tangent drawn at 3, lay off twice the length of the chord 2 1, giving the point 3′. On tangent 4, three times the length of the chord, etc. A smooth curve drawn through the points thus found will give the required curve.

NOTE.—The involute is used in developing the shape of gear teeth.

TRUE ELLIPTIC ARCH

272. To Draw a True Elliptic Arch when the Span and the Height Are Given. (Fig. 124.)

Let DE be the height and BC the span. In a supplementary diagram draw a quarter circle with radius equal to the height of the required arch. Divide this quarter circle into any number of equal parts, for example, four, giving points A, 1, 2, 3 and 4. At

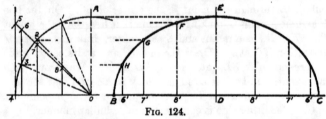

FIG. 124.

point 4 draw a line perpendicular to the base line. With O as a center and one-half the given span as a radius, draw a small arc intersecting the line drawn from 4, at the point 5. Through points 1, 2 and 3 draw perpendiculars to $O4$ cutting the line $O5$ at points 8, 7 and 6. From D, the center of the span BC, lay off distances equal to $O8$, $O7$ and $O6$, and draw perpendiculars. From points A, 1, 2 and 3 draw horizontal lines to the perpendiculars drawn from D, 8′, 7′ and 6′ respectively, giving points E, F, G and H. Through these points draw a smooth curve.

NOTE.—This curve is one-half of a true ellipse with BC as a major, and DE as a semi-minor axis.

CYCLOID

273. A cycloid is generated by a point on the circumference of a circle, when the circle rolls on a straight line. The circle is called a *generating circle* and the line on which it rolls is the *directrix*.

When the generating circle rolls on the outside circumference of another circle, as a directrix, the curve generated is called an *epicycloid*, and when the circle rolls on the inside of the circumference of another circle it is a *hypocycloid*.

274. To Draw a Cycloidal Curve Having the Span Given. (Fig. 125.)

Since the span *AB* is given, the diameter of the generating circle must be found, since the length of the line *AB* must equal the circumference of the generating circle. The required diameter can easily be found with one of the equations giving the relation between the diameter and circumference. Having found the diameter to be equal to *AC*, draw the circle with its point of tangency at *A*. Divide the circle into twelve equal parts. Number these parts as 2, 3, 4, etc. Divide the line *AB*

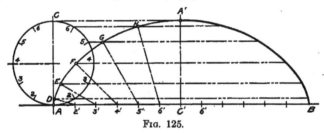

FIG. 125.

into the same number of parts as 2', 3', 4', etc. Through the points found on the circle draw lines parallel to *AB*. With 2' as a center and radius equal to the chord *A2*, draw a small arc cutting the line drawn through the point 2, giving the point *D*. With point 3' as a center and radius equal to the chord *A3*, draw a small arc on the line drawn through the point 3, giving the point *E*. The remaining points, *F, G, H* and *A'* can be similarly found, using the chordal distances *A4, A5, A6* and *AC* as radii, and 4', 5', 6' and *C'* as centers respectively. A smooth curve drawn through the points found will be the required cycloid.

NOTE.—The student will observe that the height of the curve is equal to the diameter of the generating circle.

CIRCULAR ARCS

275. To Connect Four Points by Circular Arcs. (Fig. 126.)

Let *A, B, C* and *D* be the given points. Connect the points by straight lines. Bisect each line and erect perpendiculars. On one of the perpendiculars as *EF*, assume some point as *G*. With *G* as a center and radius *GC* draw arc *BC*. From the intersection formed by perpendicular *HI* and a line connecting *G* with point *B*, draw arc *BA*, with *BJ* as a radius.   Arc *CD* may be found in the same way.

Note.—The point of tangency of arcs *AB* and *BC* lies on the line which passes through the centers from which the arcs were drawn.

276. To Connect any Number of Points with Circular Arcs. (Fig. 127.)

This problem is essentially the same as Fig. 126. Connect the points by straight lines, and erect perpendiculars and proceed as in the preceding problem.

Note.—These curves can be materially altered by assuming the center for the first arc at another point on the chosen perpendicular.

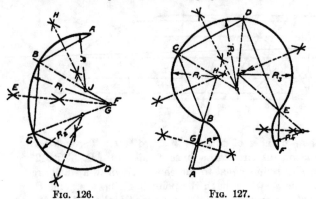

Fig. 126.   Fig. 127.

The arc connecting points *A* and *B* is dependent upon the location of its center. In this case the center was so chosen as to make the arc nearly semi-circular. This will fix the centers for all other arcs. If center *G* had been taken so as to make the radius of the arc longer, the radius *HB* would be greater, since the line passing through point *B* would assume a greater angle with chord *BC*.

277. To Draw a Circular Arc through Three Given Points when the Center is not Accessible.   (Fig. 128.)

AC is called the span and *DB* the rise. With *A* as center and *AC* as a radius draw arc *CEF*. From *C* as a center and the same radius draw arc *AGH*. From *A* and *C* draw lines through *B* intersecting the arcs at *G* and *E*. Divide arc *AG* into any number of equal parts, for example three, giving points 1 and 2.

From G lay off arcs of the same length giving points 3 and 4. Repeat this on arc CEF giving 1', 2', 3' and 4'. The intersection I of a line drawn from A to 4' with a line drawn from C to 1 will be one point on the required curve. Points J, K and L are found in a similar manner. Through these points the curve may be drawn.

NOTE.—For a true circular arc the point B must be equidistant from points A and C.

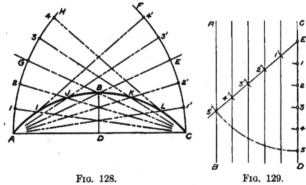

FIG. 128.   FIG. 129.

DIVIDING A SPACE

278. To Divide the Space between Two Parallel Lines into Any Number of Equal Spaces. (Fig. 129.)

Let AB and CD be parallel lines. From any point on one of the lines, as E, lay off the number of equal distances into which the space is to be divided, for example five, giving points 1, 2, 3, 4 and 5. With E as a center and radius $E5$, draw an arc intersecting line AB at the point 5'. Draw a line from E to 5'. With E as center and $E1$ as radius draw a small arc intersecting $E5'$ at 1'. Points 2', 3' and 4' may be found by spacing the length $E1'$ with the dividers on line $E5'$, or by drawing small intersecting arcs with E as a center and 2, 3 and 4 as radial distances from E. Parallel lines drawn through these points will divide the space into equal parts.

RECTIFICATION OF ARCS

279. To Rectify the Semi Circumference of a Given Circle. (Fig. 130.)

Through the center A draw the vertical and horizontal diameters BC and DE. Through C draw a line parallel to DE, indefinite in length. With D as a center and DA as a radius, draw a small arc intersecting the circumference at F. Draw line AFG. From G lay off a distance equal to three times the radius, giving point H. A line drawn from B to H will be equal to the semi-circumference of the circle. (With an error of .00006.)

NOTE.—The result obtained is close enough for all practical purposes.

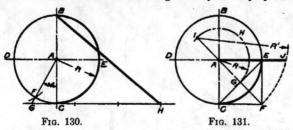

FIG. 130.   FIG. 131.

280. To Rectify the Quarter Circumference of a Given Circle. (Fig. 131.)

Through the center A draw the vertical and horizontal diameters BC and DE. Complete the square $AEFC$. Draw diagonals EC and FA. With A as a center and AG as a radius draw arc GH intersecting the line FA produced, at I. With I as a center and IF as radius, draw an arc intersecting AE produced, at J. The measure of AJ will be equal to one-quarter the circumference of the circle. (Very nearly.)

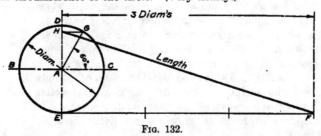

FIG. 132.

281. To Rectify the Circumference of a Given Circle. (Fig. 132.)

Through A draw horizontal and vertical diameters BC and DE. From E and parallel to BC, draw a line equal to three times the diameter of the circle giving the point F. From the center A draw a line making an angle of 60° with BC and intersecting the circumference at G. Draw GH parallel to BC.

The measure *HF* will be equal to the circumference of the circle. (Near enough for all practical purposes.)

282. To Rectify Any Short Arc of a Circle. (Fig. 133.)

Let *ABC* be the given arc and *D* its center. Draw the chord *AC* and produce it to *E*, making *CE* equal to one-half *AC*. Draw *DCF* and at *C* draw a perpendicular. (The line drawn from *C* will be tangent to the arc at the point *C*.) With *E* as center and *EA* as radius draw arc *AIJ* cutting the line *CG* at *H*. The measure of *CH* will be equal to the length of the arc *ABC*.

283. To Rectify Any Arc by Stepping Off with the Dividers. (Fig. 134.)

Let *ABC* be the given arc. At the point *A* draw a tangent *AD*. With the dividers step off any number of small dis-

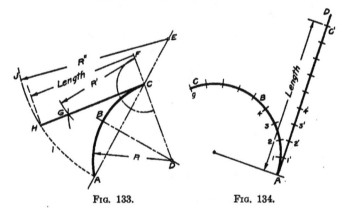

FIG. 133.   FIG. 134.

tances, giving points 1, 2, 3 and terminating at *C*. From *A* on the straight line, step off the same number of spaces giving points 1′, 2′, 3′, to *C′*. The distance from *A* to *C′* will be the measure of the arc. (Nearly.)

NOTE.—This is the method commonly used by draftsmen, and for most work is accurate enough.

TANGENT LINES

284. To Draw a Line Tangent at a Given Point on a Circular Arc when its Center is not Accessible. (Fig. 135.)

With the given point *A* as a center and any convenient radius, draw small arcs intersecting the given arc at points

B and C. Draw the chord BC. From the point A draw a line parallel to the chord, using any method. The line AD will be the required tangent.

285. To Draw Lines Tangent to a Given Circle from a Point Outside. (Fig. 136.)

Through the point A, the given point, and the center of the circle C, draw the line AB. Bisect the line AC. With D,

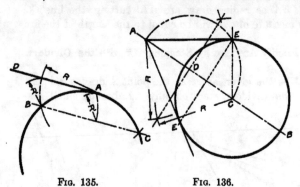

Fig. 135.   Fig. 136.

the middle point of AC, as a center and DC as a radius, draw a semicircle intersecting the given circle at point E. A line drawn from A to E will be tangent at the point E.

Note.—Angle AEC is a right angle, and from geometry we have, "Any angle inscribed in a semicircle is a right angle." Since AE is perpendicular to the radius CE, it is tangent to the circle at the point E.

TO TRANSFER POLYGONS

286. To Transfer a Polygon Having Four Sides, to a New Location. (Fig. 137.)

Let $ABCD$ be the given polygon, and let $A'B'$ be the new base to which the polygon is to be transferred. With A' and B' as centers, also BC and AC as radii, respectively, draw small arcs, the intersection of which will give C'. With A' and C' as centers, also AD and CD as radii respectively, draw small arcs, the intersection of which will give D'. Join the points found.

287. To Transfer a Polygon Having More than Four Sides, to a New Location. (Fig. 138.)

Let $A'B'$ be the given base to which the polygon is to be transferred. Points C' and D' are found as in Fig. 137.

Point E' is found by drawing small arcs from D' and A' with DE and AE as radii. Point F' may be located by intersecting arcs drawn from A' and E' as centers, with AF and EF as radii respectively.

NOTE.—By this method a polygon of any number of sides may be transferred to a new location.

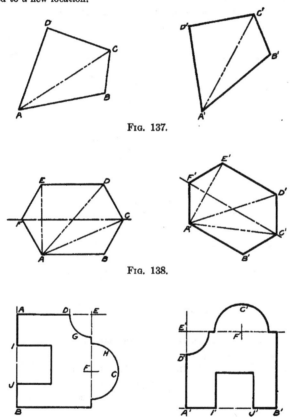

FIG. 137.

FIG. 138.

FIG. 139.

288. To Transfer any Figure which can be Bounded by Rectangular Lines. (Fig. 139.)

Let ABC be the given figure which is to be transferred, and turned through an angle of 90°, to $A'B'$ as a base. At A'

erect a perpendicular and with the dividers step off $A'D'$ and $A'E'$, equal to AD and AE. Then E' is the center of the arc DG. The vertical distances on line EF are laid off on the horizontal line $E'F'$. Then F' is the center of arc CH. Points I' and J' are laid off with the measurements obtained by measuring from A to I, and A to J, etc.

289. To Construct any Symmetrical Polygon of Curved Lines Having One Side Given. (Fig. 140.)

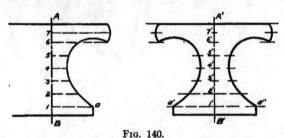

Fig. 140.

On the given center line $A'B'$ lay off any convenient number of distances as $1'$, $2'$, etc., and through these points draw lines parallel to the base. With the dividers, from $1'$, and with distance equal to $1a$, lay off $1'a'$ and $1'a''$, thus giving one point on each side of the center line. The other points may be found in a similar manner. Connect the points thus found by smooth curves.

CONIC SECTIONS

290. A conic section is a plane curve formed by the line of intersection of a cone and a plane.

Figs. 141 and 142 are illustrations of a double, right circular cone and a number of planes in various positions.

The cutting plane may be passed in seven different positions with respect to the cone.

First.—If the plane be passed through the vertex and perpendicular to the axis, it will produce a *point*.

Second.—When the plane coincides with an element, it will produce a *line*.

Third.—If the plane be passed through the axis, it will produce *two intersecting straight lines*.

Fourth.—When a plane is passed through the cone perpendicular to the axis, a *circle* will be the result.

Fifth.—If a plane be passed oblique to the axis and making an angle greater than the elements, an *ellipse* will be formed.

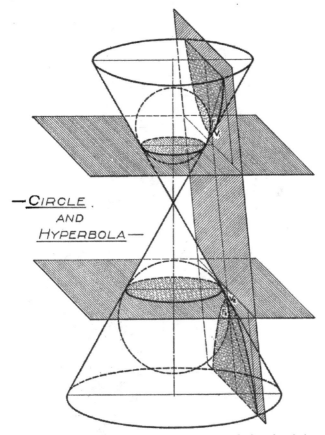

Fig. 141.—Showing a cone and cutting planes, producing the circle and hyperbola.

Sixth.—A *parabola* is produced by passing a plane parallel to an element.

Seventh.—If a plane be passed making a smaller angle with the axis, than the elements, an *hyperbola* will be formed.

The first three cases are not generally considered as conic sections.

The true conic sections are the *Circle, Ellipse, Parabola,* and *Hyperbola.*

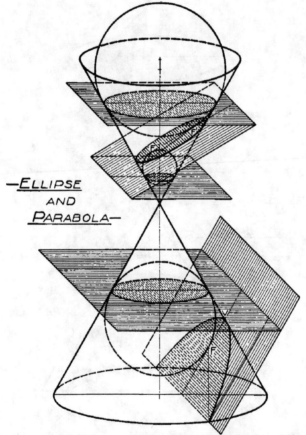

—ELLIPSE
AND
PARABOLA—

FIG. 142.—Showing a cone and cutting planes, producing the ellipse and parabola.

By referring to the figures it will be seen how these curves are formed by the aid of a cone and a plane.

NOTE.—A *complete cone* lies as much above as below the vertex. The two parts are known as the lower and upper *nappes.*

291. Ellipse.

If a plane is passed through a cone or cylinder perpendicular to the axis, the line of intersection will be a circle. If, however, the plane is passed oblique to the axis of the cone or cylinder the line of intersection will be an ellipse.

292. An ellipse is a closed curve which has two axes, namely, a *Major Axis*, or *Longest Diameter*, and a *Minor Axis*, or *Shortest Diameter*. It also has two points on the major axis called the *Foci*, which lie equidistant from its center. In nearly all cases where an elliptical curve is wanted, the dimensions of the long and short axes are known. Next to the circle the ellipse is most commonly met with in mechanical drawing.

293. The ellipse may be defined as being a closed curve generated by a point moving in a plane in such a way that its distance measured from two points on the major axis, called the foci, is a constant. For illustration, see Fig. 143. If a piece of thread were fastened to pins placed at the foci F and F', of such length that if drawn taut with the point of a pencil, the pencil would touch the point P, then the sum of the distances PF and PF' would be the constant referred to. If the pencil be now moved and the thread be kept taut, the curve described by the pencil will be an ellipse.

There are a number of methods by which an ellipse may be drawn. With the exception of the first method, to be explained, they all depend upon the location of a number of points through which a curved line is to be drawn with the irregular curve. This in many cases will not lead to good results, especially if the ellipse is small, say under 3 inches long. To offset this difficulty many ingenious methods have been devised to draw an ellipse with circular arcs. These methods will not give a true elliptic curve, but are accurate enough in nearly all cases where such a curve is wanted.

294. To Draw an Ellipse.

First Method.—(Fig. 143.) By intersecting arcs.

Having given the length of the major and minor axes AB and CD respectively. Find the foci by drawing small arcs with AO as a radius and C as a center, cutting AB in F and F'. With F as a center and any radius greater than AF, say AG, describe a small arc. From F' as a center and GB as a radius, describe a small arc intersecting the first arc in H. This will be one point on the required curve. Again, with F as a center and AJ as a radius, draw another arc; and with F' as a center and JB as a radius draw an arc intersecting the first arc at K, which will be another point on the curve. Similarly, other points may be found, through which a smooth curve may be drawn.

NOTE.—$FC + CF' = AB$, also $FH + HF' = AB$, etc.

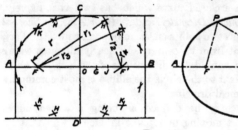

FIG. 143.

Second Method—(Fig. 144.)   By the *trammel* method.

Having giving the length of the major and minor axes intersecting at the center O.   On a narrow strip of drawing paper, a little longer than half the major axis, mark off the points J, I and P, making JP equal to the semi-major, and IP equal to the semi-minor axes.   Now place the paper strip so that J becomes J' on

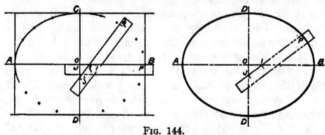

FIG. 144.

the minor axis, and I becomes I' on the major axis.   Place a point at P' which will be one point on the required curve.   Similarly, other points may be found by placing the strip in different positions.   Draw a smooth curve through the points.

Third Method.—(Fig. 145.)   With intersecting lines.

Having given the major and minor axes AB and CD respectively.   Through points A and B draw vertical lines and through points C and D draw horizontal lines, forming the rectangle $EFHG$.   Divide AE and AO, each into the same number of equal

parts. Transfer these divisions to the other lines, as illustrated
in the figure. For the upper right-hand part of the curve,
draw lines from C to points 1, 2, 3 and 4. From D draw lines
through points 5, 6, 7 and 8, intersecting the lines drawn from C
in points as shown. Draw a smooth curve through the points.

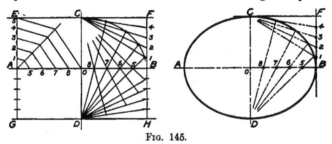

Fig. 145.

Fourth Method.—(Fig. 146.) An approximate ellipse, using
four circular arcs.

Having given the major and minor axes AB and CD intersecting at O. On the minor axes lay off OE and OF equal to the
difference between the major and minor axes. Divide OE into
four equal parts. Lay off OG and OH on the major axis, each

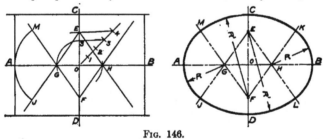

Fig. 146.

equal to three-fourths of OE. Through the points E, G, F and
H draw lines as shown. With G as a center and GA as radius,
draw an arc intersecting lines FG and EG produced, at points
M and J; with the same radius and H as a center draw arc KL.
With E and F as centers, draw arcs JL and MK, respectively.

Note.—This is an approximate ellipse sufficiently accurate for most purposes in technical drawing, but will not apply when the length of the minor
axis is less than half the length of the major axis.

The points of tangency, for good results, must be exactly on the lines drawn
through the centers E, G, F and H.

295. To Draw an Elliptic Arc, Having Given the Length of a Chord and One Axis. (Fig. 147.)

Let AB be the given axis, CD the chord and O the intersection of the axis and chord. With the chord CD as one side and AO as the height, draw the rectangle $EFDC$. Divide EC and CO, each into the same number of equal parts. · From A draw lines to the points found on CE. From B draw lines passing through the points on CO, intersecting the lines drawn from A at points H, I and J. Find similar points for the other half. Through these points draw a smooth curve, thereby completing the entire elliptic arc required.

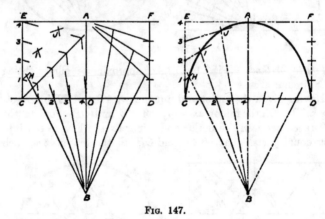

Fig. 147.

296. Parabola.

The parabola is a curve formed by the intersection of the surface of a cone with a plane taken parallel to one of its elements.

297. To Draw a Parabolic Curve, Having the Height and Horizontal Length Given as Limits. (Fig. 148.)

Let AB be the height and BC the length of a rectangle which represents the limits of the curve of which DE is the axis.

Divide AB and BE, each into the same number of equal parts. From the points on BE draw lines parallel to the axis DE. From D draw lines to the points found on line AB. The intersection of these lines will determine points 4, 5, and 6 which are points on the required curve. Since the distance from point 6 to B is somewhat long, it will be well to find another point on the curve. This can be found by bisecting B3 and drawing lines 7 as before, thus locating point 8. Through these points the curve may be drawn. In a similar manner the other side may be found.

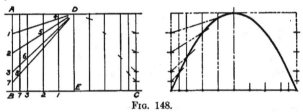

FIG. 148.

298. Hyperbola.

The hyperbolic curve is formed if the cutting plane makes an angle with the base, which is greater than an element of the cone.

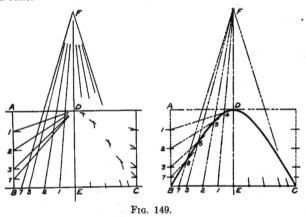

FIG. 149.

299. To Draw an Hyperbolic Curve within Given Limits. (Fig. 149.)

This curve, unlike the two former curves, may vary considerably within its limits, unless the location of the point F, called the focus, is given.

Let AB be the height and BC the length of the rectangle which represents the limits of the curve. Also let a line drawn perpendicular from the center of BC be the axis. The point F, or focus, may be taken at any convenient point on the axis.

As in the previous two problems the lines *AB* and *BE* are each divided into an equal number of parts, and lines are drawn to points *D* and *F* whose intersections give points 4, 5, 6 and 8 as before, through which the curve may be drawn.

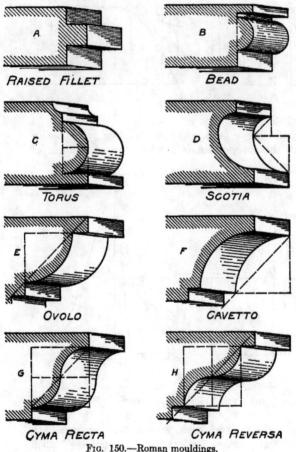

RAISED FILLET   BEAD

TORUS   SCOTIA

OVOLO   CAVETTO

CYMA RECTA   CYMA REVERSA

FIG. 150.—Roman mouldings.

ROMAN MOULDINGS

300. The mouldings here shown are composed of arcs of circles and straight lines, and all the constructing angles are drawn at 45°.

Fig. 150.—*A* is a *Raised Fillet*. Its projection is generally made equal to its height. It may also be a depression, in which case it is called a *Sunk Fillet*.

B shows a *Bead* which is a moulding consisting of a semicircle on a flat surface. This is also called a *Half Round*.

C is a *Torus* which is composed of a semicircle and a fillet. The semicircle projects from the fillet an amount equal to its radius. The concave addition above the fillet and the projection below the half round shows a combination of mouldings used in the base of a *Doric Column*.

D illustrates the *Scotia* which consists of two quarter circles one of which has a radius equal to twice the radius of the other. The centers of both arcs must be on the same horizontal line. It is generally applied in the bases of columns.

E shows the *Ovolo* which is composed of a quarter of a circle and an upper and lower fillet. Without the addition of the fillet it is called a *Quarter Round*. Its construction is made apparent by referring to the figure.

F is the *Cavetto* which, like the ovolo, consists of a quarter circle with a sunk fillet below. By referring to the figure it will be seen that it is exactly the reverse of the ovolo. The center for describing the quarter circle is without instead of within the moulding. It is also called a *Concave Moulding*.

G shows the *Cyma Recta* which is a moulding of double curvature and two fillets. The curve is composed of quarter-circles; the upper or concave portion of the curve has its center without, while the lower curve has its center within the moulding. Both centers are on the same horizontal line. This moulding is frequently called an *Ogee*.

H illustrates the *Cyma Reversa* which, like the cyma recta, is composed of two quarter circles and an upper and lower fillet. It is distinguished from the former by having its convex part above and the concave part below a horizontal center line.

NOTE.—The drawing of these mouldings is so apparent by referring to the illustrations that further description is unnecessary.

These mouldings in simple outline or in combination may be purchased from almost any dealer in building materials.

CHAPTER IX

MENSURATION

301. Introductory.

Mensuration, as indicated in the introduction, is a valuable auxiliary to mechanical drawing. In the making of working drawings or designs, the draftsman is frequently required to make calculations of areas or circumferences of circles, or of volumes and weights, or of the amount of material required for a specific project. The essentials for such calculations comprise the following chapter. This chapter is not intended as a course in mathematics. It offers merely a selection of definitions and equations of mensuration that are of importance in the practice of mechanical drawing. The formulas are stated and each one is illustrated, by one or more examples, showing how they are used, but the derivation of these formulas belongs to pure mathematics.

From time to time the student in mechanical drawing should make, in connection with his specific problem in drawing, such calculations of areas, amounts of materials, volumes, weights and costs, as would be made in actual drawing-room practice. This will give him the required facility in the use of mathematical formulas and equations.

ANGLES

302. The divergence of any two lines from their point of intersection is called an angle.

The vertex of an angle is the point of intersection of the two lines which are called the sides of the angle.

Any angle may be measured in degrees or in radians.

The unit of circular measure is the degree, which is 1/360th of a circle.

A degree may be divided into 60 minutes and a minute into 60 seconds. An angle of 35 degrees, 54 minutes and 13 seconds is written 35° 54′ 13″.

A circle contains 2π radians and one radian is called a unit angle which is found by dividing 360 by 2π.

Thus, $$\frac{360°}{2\pi} = \frac{360°}{2 \times 3.1416} = 57°.3 \text{ nearly.}$$

Two angles whose sum equals a right angle or 90°, are said to be complementary to each other. If their sum equals two right angles or 180°, they are supplementary to each other.

For illustrations of the following definitions see Fig. 151.

303. Right Angle.

A right angle is one of 90°. It is formed when drawing a line perpendicular to a given line.

304. Acute Angle.

An acute angle is less than a right angle. It may be any angle between 0° and 90°.

305. Obtuse Angle.

An obtuse angle is greater than a right angle and less than two right angles. It may be any angle between 90° and 180°.

TRIANGLES

306. A triangle is any plane figure bounded by three straight lines.

The sum of the three angles of any triangle is equal to 180°. If one of the angles is 90°, or a right angle, the sum of the other two is 90°.

In radian measure, the sum of the three angles is equal to π radians.

The perpendicular distance, from the base to the vertex, is called the altitude or height. The area is equal to the length of the base multiplied by one half the altitude or perpendicular.

For illustrations of the following triangles see Fig. 151.

307. Equilateral Triangle.

In an equilateral triangle the three sides are all of equal length and each angle is 60°.

$$\text{Area} = A = \frac{bh}{2}, \ h = \frac{b\sqrt{3}}{2}, \ b = \frac{2A}{h}$$

Example.—Find the area when b = 4 inches.

Since $\qquad h = \frac{b\sqrt{3}}{2} = \frac{4 \times 1.732}{2} = 3.464$ inches,

then $\qquad A = \frac{bh}{2} = \frac{4 \times 3.464}{2} = 6.93$ square inches.

If the area is known, the length of the sides may be found.

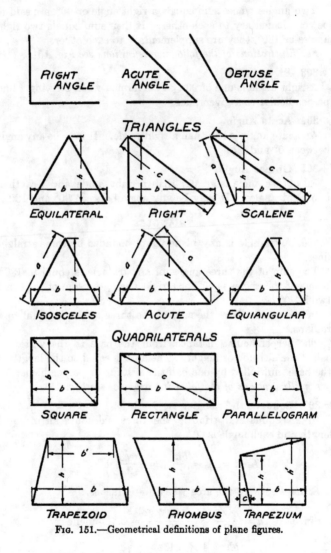

FIG. 151.—Geometrical definitions of plane figures.

Example.—Let $A = 24$ square inches. Find b.

Since
$$A = 24 = \frac{bh}{2} = \frac{b^2\sqrt{3}}{4} = .433b^2$$

then
$$b = \sqrt{\frac{24}{.433}} = \sqrt{55.427} = 7.44 \text{ inches.}$$

308. Right Angle Triangle.

A right angle triangle is one having two sides meeting at an angle of 90°.

$$\text{Area} = A = \frac{bh}{2}$$

$$h = \sqrt{c^2 - b^2}, \quad c = \sqrt{b^2 + h^2}, \quad b = \sqrt{c^2 - h^2}$$

Examples.—Find the area, and c, when $b = 6$ inches and $h = 4$ inches.

Thus,
$$A = \frac{bh}{2} = \frac{6 \times 4}{2} = \frac{24}{2} = 12 \text{ square inches,}$$

and
$$c = \sqrt{b^2 + h^2} = \sqrt{36 + 16} = \sqrt{52} = 7.211 \text{ inches.}$$

If c and h were known to be 9 and 6 inches, respectively, then

$$b = \sqrt{c^2 - h^2} = \sqrt{81 - 36} = \sqrt{45} = 6.708 \text{ inches.}$$

If c and b were known to be 15 and 12 inches, respectively, then

$$h = \sqrt{c^2 - b^2} = \sqrt{225 - 144} = \sqrt{81} = 9 \text{ inches.}$$

309. Scalene Triangle.

This is a triangle having no two sides or angles equal.

$$\text{Area} = A = \frac{bh}{2}, \quad h = \sqrt{a^2 - \left(\frac{c^2 - a^2 - b^2}{2b}\right)^2}$$

Example.—Let $a = 5$, $b = 7$ and $c = 11$ inches.

Then

$$A = \frac{bh}{2} = \frac{b}{2}\sqrt{a^2 - \left(\frac{c^2 - a^2 - b^2}{2b}\right)^2}$$

$$= \frac{7}{2}\sqrt{5^2 - \left(\frac{11^2 - 5^2 - 7^2}{2 \times 7}\right)^2}$$

$$= 3.5 \sqrt{25 - \left(\frac{121 - 25 - 49}{14}\right)^2}$$

$$= 3.5 \sqrt{25 - \left(\frac{47}{14}\right)^2} = 3.5 \sqrt{25 - 3.357^2}$$

$$= 3.5 \sqrt{13.730} = 3.5 \times 3.705 = 12.96 \text{ square inches.}$$

NOTE.—The area may be found by first finding a numerical value for h and substituting this in the first equation.

Example.—Let $a = 6$, $b = 9$ and $c = 12$ inches.

Since

$$h = \sqrt{a^2 - \left(\frac{c^2 - a^2 - b^2}{2b}\right)^2} = \sqrt{6^2 - \left(\frac{12^2 - 6^2 - 9^2}{2 \times 9}\right)^2}$$

$$= \sqrt{36 - \left(\frac{144 - 36 - 81}{18}\right)^2} = \sqrt{36 - \left(\frac{27}{18}\right)^2}$$

$$= \sqrt{36 - 1.5^2} = \sqrt{36 - 2.25} = \sqrt{33.75} = 5.81 \text{ inches,}$$

then

$$A = \frac{bh}{2} = \frac{9 \times 5.81}{2} = 4.5 \times 5.81 = 26.14 \text{ square inches.}$$

310. Isosceles Triangle.

This triangle has two equal sides, hence two equal angles.

$$\text{Area} = A = \frac{bh}{2}, \quad h = \sqrt{a^2 - \left(\frac{b}{2}\right)^2}$$

Example.—Let $a = 10$ and $b = 8$ inches.

Since

$$h = \sqrt{a^2 - \left(\frac{b}{2}\right)^2} = \sqrt{10^2 - \left(\frac{8}{2}\right)^2}$$

$$= \sqrt{100 - 4^2} = \sqrt{100 - 16} = 9.165 \text{ inches,}$$

then

$$A = \frac{bh}{2} = \frac{8 \times 9.165}{2} = 4 \times 9.165 = 36.66 \text{ square inches.}$$

311. Acute Triangle.

An acute triangle is one in which each angle is less than 90°.

$$\text{Area} = A = \frac{bh}{2} = \frac{b}{2}\sqrt{c^2 - \left(\frac{c^2 + b^2 - a^2}{2b}\right)^2}$$

Example.—Let $a = 8$, $b = 12$ and $c = 10$ inches.

$$\text{Since } h = \sqrt{c^2 - \left(\frac{c^2 + b^2 - a^2}{2b}\right)^2} = \sqrt{10^2 - \left(\frac{10^2 + 12^2 - 8^2}{2 \times 12}\right)^2}$$

$$= \sqrt{100 - \left(\frac{100 + 144 - 64}{24}\right)^2} = \sqrt{100 - \left(\frac{180}{24}\right)^2}$$

$$= \sqrt{100 - 7.5^2} = \sqrt{43.75} = 6.614 \text{ inches,}$$

$$\text{then } A = \frac{bh}{2} = \frac{12 \times 6.614}{2} = 39.68 \text{ square inches.}$$

312. Equiangular Triangle.

This is a triangle in which the three angles are equal. Since the sum of the three angles is equal to 180°, each angle will be equal to 60°.

An equiangular triangle is also equilateral, therefore the equations for the latter will answer for the former.

313. Area of any Triangle.

If the length of the sides of any triangle is known, but not the altitude, the area may be found from the following equation:

$$\text{Area} = A = \sqrt{S(S - a)\ (S - b)\ (S - c)},$$

where S equals one-half the length of the perimeter.

$$\text{Hence} \qquad S = \frac{a + b + c}{2}$$

Example.—Let $a = 5$, $b = 7$ and $c = 10$ inches.

$$\text{Since } S = \frac{a + b + c}{2} = \frac{5 + 7 + 10}{2} = \frac{22}{2} = 11 \text{ inches,}$$

$$\text{then } A = \sqrt{S(S - a)\ (S - b)\ (S - c)}$$

$$= \sqrt{11(11 - 5) \times (11 - 7) \times (11 - 10)}$$

$$= \sqrt{11 \times 6 \times 4 \times 1} = \sqrt{264} = 16.24 \text{ square inches.}$$

QUADRILATERALS

314. A quadrilateral is any plane figure bounded by four straight lines.

The diagonal of a quadrilateral is the straight line connecting opposite angles.

The altitude is the perpendicular height of the figure.

The sum of the four angles of any quadrilateral is equal to 360°.

For illustrations of the following quadrilaterals see Fig. 151.

315. Square.

A square is a plane figure having four sides of equal length, all its angles being right angles.

$$\text{Area} = A = b^2, \text{ or } \frac{c^2}{2}$$

$$\text{Length of side} = b = \sqrt{A}$$

$$\text{Length of diagonal} = c = \sqrt{2A} = \sqrt{2b^2} = 1.414\sqrt{b^2}$$

Example.—Let $b = 9$ inches. Find the area.

Thus, $A = b^2 = 9^2 = 81$ square inches.

If the area is known to be 165 square inches, find the length of one side; also the length of a diagonal.

Thus, $b = \sqrt{A} = \sqrt{165} = 12.845$ inches,

and $c = \sqrt{2A} = \sqrt{2 \times 165} = \sqrt{330} = 18.16$ inches.

316. Rectangle.

The rectangle is a plane figure having opposite sides parallel, whose angles are right angles.

$$\text{Area} = A = bh$$

$$b = \frac{A}{h} = \sqrt{c^2 - h^2}$$

$$h = \frac{A}{b} = \sqrt{c^2 - b^2}$$

$$c^2 = b^2 + h^2, \quad c = \sqrt{b^2 + h^2}$$

Example.—Let $b = 20$ and $h = 16$ inches.

Then $A = bh = 20 \times 16 = 320$ square inches.

If $A = 154$ square inches and $h = 13$ inches,

then $b = \frac{A}{h} = \frac{154}{13} = 11.846$ inches.

317. Parallelogram.

A parallelogram is a plane figure having its opposite sides parallel and its opposite angles equal.

$$\text{Area} = A = bh, \quad b = \frac{A}{h}, \quad h = \frac{A}{b}$$

Example.—If $b = 14$ and $h = 17.5$ inches,

then $\qquad A = bh = 14 \times 17.5 = 245$ square inches.

If the area were given as 38.5 square inches and the length 7.65 inches,

then $\qquad h = \dfrac{A}{b} = \dfrac{38.5}{7.65} = 5.033$ inches.

318. Trapezoid.

The trapezoid is a plane figure having two sides parallel. Two of its angles may be right angles.

$$\text{Area} = A = \frac{b + b'}{2} \times h = \frac{h(b + b')}{2}$$

Example.—Let $b = 23$, $b' = 16$, and $h = 12$ inches.

Then

$$A = \frac{h(b + b')}{2} = \frac{12(23 + 16)}{2} = 6 \times 39 = 234 \text{ square inches.}$$

If in the above example the area and the length of its sides were given, to find the height:

Since $\quad A = \dfrac{h(b + b')}{2}, \quad 2A = h(b + b'),$

then $\qquad h = \dfrac{2A}{b + b'} = \dfrac{2 \times 234}{23 + 16} = \dfrac{468}{39} = 12$ inches.

319. Rhombus.

The rhombus is a plane figure having equal sides, whose angles are not right angles.

$$\text{Area} = A = bh, \quad b = \frac{A}{h}, \quad h = \frac{A}{b}$$

The calculations of area and length of sides are the same as for the parallelogram.

320. Trapezium.

This is a plane figure with no two sides parallel. One of its angles may be a right angle, or two opposite angles may be equal.

$$\text{Area} = A = b\left(\frac{h + h'}{2}\right) - \frac{hc}{2} = \frac{b(h + h') - hc}{2}$$

Example.—Let $b = 10.75$, $c = 3.5$, $h = 8.25$ and $h' = 12$ inches.

Then

$$A = \frac{b(h + h') - hc}{2}$$

$$= \frac{10.75\,(8.25 + 12) - 8.25 \times 3.5}{2}$$

$$= \frac{10.75 \times 20.25 - 8.25 \times 3.5}{2}$$

$$= \frac{217.6875 - 28.875}{2} = \frac{188.812}{2}$$

$$= 94.40 \text{ square inches.}$$

CIRCLE

321. The circumference of a circle is a curved line, all points of which are equally distant from a point within, called the center.

The circle itself is the space within the circumference; this is called its area.

If two or more circles are drawn from a common center they are said to be concentric.

The area contained between two concentric circles is called an annulus.

Let C = circumference, R, or r = radius,
       A = area, D, or d = diameter.

For illustrations see Fig. 152.

322. Circumference.

$$C = \pi D = 3.1416D, \text{ or } \frac{22}{7} D$$

also

$$C = 2\pi R = 6.2832R, \text{ or } \frac{44}{7} R$$

Example.—Find the circumference of a circle whose diameter is 13 inches.

Thus, $$C = \frac{22}{7} D = \frac{22}{7} \times 13 = 40.84 \text{ inches.}$$

323. Area.

$$A = \frac{\pi d^2}{4} = .7854d^2, \text{ or } \frac{22}{7} \times \frac{d^2}{4} = \frac{11d^2}{14}$$

also

$$A = \pi r^2 = 3.1416r^2, \text{ or } \frac{22}{7} r^2$$

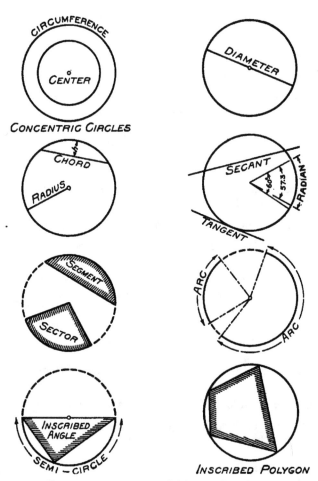

Fɪɢ. 152.—Geometrical definitions of plane figures.

Example.—Find the area of a circle whose diameter equals 18 inches.

Thus, $A = \dfrac{\pi d^2}{4} = \dfrac{3.1416 \times 18 \times 18}{4} = 254.47$ square inches

or $\qquad A = \pi r^2 = 3.1416 \times \left(\dfrac{18}{2}\right)^2 = 254.47$ square inches.

324. Area of Annulus.

A = area, D = large diameter, d = small diameter,
R = large radius, r = small radius.

$$A = \frac{\pi D^2}{4} - \frac{\pi d^2}{4} = \frac{\pi}{4}(D^2 - d^2) = .7854(D + d)(D - d)$$

or $\quad A = \pi R^2 - \pi r^2 = \pi(R^2 - r^2) = 3.1416(R + r)(R - r)$

Example.—Find the area of an annulus whose large diameter is 14 and small diameter 9 inches.

Thus, $A = .7854(D + d)(D - d) = .7854(14 + 9)(14 - 9)$

$\qquad\qquad = .7854 \times 23 \times 5 = 90.32$ square inches.

325. Diameter and Radius.

Every straight line which passes through the center of a circle and terminates at the circumference, is called a diameter.

The diameter of a circle may be found by transposing the equation:

$$C = \pi d, \quad \text{then } d = \frac{C}{\pi} = .318C$$

Example.—Find the diameter of a circle whose circumference is 36 inches.

Thus, $\quad d = .318C = .318 \times 36 = 11.448$ inches.

If the area of a circle is given as 182 square inches, the diameter may be found from the equation:

$$A = .7854d^2, \quad \text{then } d^2 = \frac{A}{.7854} \text{ or } d = \sqrt{\frac{A}{.7854}}$$

Example.—Find the diameter of a circle whose area equals 495 square inches.

Thus,

$$d = \sqrt{\frac{A}{.7854}} = \sqrt{\frac{495}{.7854}} = \sqrt{630.252} = 25.10 \text{ inches.}$$

A radius is any line drawn from the center to the circumference of a circle.

The radius may be found from the circumference or area equations.

$$C = 2\pi R \quad \therefore R = \frac{C}{2\pi} = \frac{C}{6.2832} \doteq .159C.$$

Example.—Find the radius of a circle whose circumference is 122 inches.

Thus, $R = .159C = .159 \times 122 \doteq 19.39$ inches.

Example.—Find the radius of a circle whose area is 220 square inches.

Since $\qquad A = \pi R^2, \; R = \sqrt{\dfrac{A}{\pi}} = .564\sqrt{A}$

then $\qquad R = .564\sqrt{A} = .564\sqrt{220}$

$$= .564 \times 14.832 = 8.365 \text{ inches.}$$

326. Chord.

The chord of a circle is a straight line joining any two points of the circumference. When the chord passes through the center, it becomes a diameter.

The length of a chord may be found, when the radius of the circle and the perpendicular distance from the center of the chord to the circumference are known, from the formula:

$$L = 2\sqrt{h(2R - h)}, \text{ or } 2\sqrt{h(D - h)}$$

where $\qquad\qquad L =$ length of chord
and $\qquad\qquad h =$ perpendicular height.

Example.—Find the length of the chord of a circle when R equals 12 inches and h equals 4 inches.

Thus, $\qquad L = 2\sqrt{h(2R - h)} = 2\sqrt{4(2 \times 12 - 4)}$

$$= 2\sqrt{4 \times (24 - 4)} = 2\sqrt{4 \times 20}$$

$$= 2\sqrt{80} = 2 \times 8.94 = 17.88 \text{ inches.}$$

327. Tangent Lines.

A tangent is a straight line which touches the circumference at one point only, and will not intersect the circle if produced.

Any number of tangents may be drawn to a circle.

Two circumferences are tangent when they touch at only one point.

328. Secant.

A secant is a straight line which intersects the circumference of a circle in two points.

329. Segment.

A segment is a plane figure bounded by the arc of a circle and a chord. To one chord there are always two arcs, therefore, two segments; the smaller of the two is always understood unless the contrary is stated.

If the length of the chord and the length of h are known, the area of a segment may be found very nearly from the formula:

$$\text{Area} = A = \frac{h}{6L}(3h^2 + 4L^2).$$

Example.—Let $h = 3$ and $L = 12$ inches.

$$\text{Then } A = \frac{h}{6L}(3h^2 + 4L^2) = \frac{3}{6 \times 12}\left[3 \times (3)^2 + 4 \times (12)^2\right]$$

$$= \frac{3}{72}(3 \times 9 + 4 \times 144) = \frac{3}{72}(27 + 576)$$

$$= \frac{3}{72} \times 603 = \frac{1809}{72} = 25.12 \text{ square inches.}$$

330. Sector.

A sector is the area bounded by the arc of a circle and two radii.

If the angle and the radius of the sector are given, the area may be found from the formula:

$$\text{Area} = A = \frac{\pi R^2 N}{360} = \frac{R^2 N}{114.5} = .0087\, R^2 N$$

where N = the number of degrees.

Example.—Find the area of a sector when $N = 54°$ and $R = 7$ inches.

Thus,

$$A = .0087\, R^2 N = .0087 \times 7 \times 7 \times 54 = 23.02 \text{ square inches.}$$

If the length of the arc and the radius are known the area may be found from:

$$\text{Area} = A = L\frac{R}{2} = \frac{LR}{2}$$

where L = the length of arc, in inches.

Example.—Find the area of a sector when $L = 6.5$ and $R = 7$ inches.

Thus,

$$A = \frac{LR}{2} = \frac{6.5 \times 7}{2} = 6.5 \times 3.5 = 22.75 \text{ square inches.}$$

331. Length of Circular Arc.

The length of a circular arc may be any portion of the circumference of a circle.

The length of the circular arc for a complete circumference may be expressed as:

$$\text{Length of arc} = L = 2\pi R, \text{ for } 360 \text{ degrees.}$$

Then the length of arc of one degree $= \dfrac{2\pi R}{360}$

and for N degrees, $L = \dfrac{2\pi RN}{360} = \dfrac{\pi}{180} RN = .0174 RN$

Example.—Find the length of the arc of a circle when $R = 9$ inches and $N = 74$ degrees.

Thus,

$$L = .0174RN = .0174 \times 9 \times 74 = 11.58 \text{ inches.}$$

Example.—Find the length of the arc of a circle whose radius is 7.5 inches and the number of degrees subtended by the arc is 67° 15′.

NOTE.—15 minutes is equal to .25 of one degree.

Thus,

$$L = .0174RN = .0174 \times 7.5 \times 67.25 = 8.77 \text{ inches.}$$

332. Inscribed Angle.

An inscribed angle is one whose vertex and the extremities of whose sides touch the circumference of a circle.

If the angle be inscribed in a semi-circle, it is a right angle.

333. Inscribed Polygon.

A polygon is the area of a plane figure bounded by three or more straight lines. If there are three sides it is a triangle; if four sides it is a quadrilateral; if five sides it is a pentagon, and with six sides it is a hexagon.

An inscribed polygon is one which has all its vertices in the circumference of a circle.

SOLIDS

334. A solid is any figure whose elements are not all in the same plane. It has three dimensions: length, breadth, and height.

335. Prism.

A prism is a solid bounded by a number of parallelograms which terminate at both ends by parallel and equal polygons or bases. If the parallelograms are perpendicular to the base, it is called a right prism; if inclined, it is called an oblique or scalene prism.

The altitude of a prism is the perpendicular distance between its bases.

A regular prism is one whose bases are regular polygons and whose axis is perpendicular.

The volume of any right or oblique prism is equal to the product of the area of one base multiplied by the altitude.

The lateral surface of any prism is equal to the length of its perimeter times the altitude.

336. Abbreviations.

A = area of base, V = volume, S = lateral surface,
S' = total surface, W = weight, L = slant height.

For illustrations of solid figures see Fig. 153.

337. Square Prism.

(A)

$$V = a^2h, \quad h = \frac{V}{a^2}, \quad a = \sqrt{\frac{V}{h}}$$

Examples.—Find the volume and weight of a square prism, the length of whose sides is 3 inches and height 5 inches.

Thus, $\quad V = a^2h = 3 \times 3 \times 5 = 45$ cubic inches.

If cast iron weighs .26 per cubic inch its weight will be,

$$W = .26a^2h = .26 \times 45 = 11.70 \text{ pounds.}$$

Find the height if the volume is 16 cubic inches and the length of its base is 2.25 inches.

Thus, $\quad h = \dfrac{V}{a^2} = \dfrac{16}{2.25 \times 2.25} = \dfrac{16}{5.0625} = 3.16$ inches.

If $\quad V = 38$ cubic inches and $h = 6$ inches, find a.

Thus, $\quad a = \sqrt{\dfrac{V}{h}} = \sqrt{\dfrac{38}{6}} = \sqrt{6.333} = 2.51$ inches.

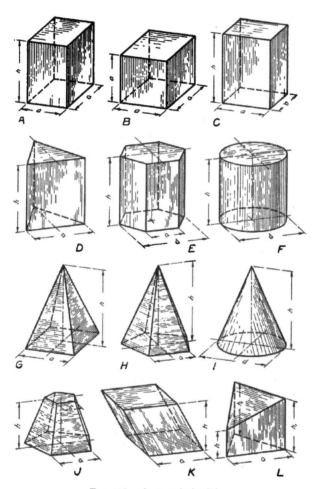

Fig. 153.—Geometrical solids.

(B)

338. Cube.

$$V = a^3, \ a = \sqrt[3]{V}$$

Examples.—Find the volume, surface and weight of a cube with 4 inch faces.

Thus,

$$V = a^3 = 4 \times 4 \times 4 = 64 \text{ cubic inches.}$$

$$S = 4\,a^2 = 4 \times 4 \times 4 = 64 \text{ square inches.}$$

$$S' = 6a^2 = 6 \times 4 \times 4 = 96 \text{ square inches.}$$

If lead weighs .41 pound per cubic inch, the weight will be,

$$W = .41\,a^3 = .41 \times 64 = 26.24 \text{ pounds.}$$

If the volume of a cube is 60 cubic inches, find the length of its edges.

Thus,

$$a = \sqrt[3]{V} = \sqrt[3]{60} = 3.91 \text{ inches.}$$

339. Rectangular Prism.

(C)

$$V = abh, \quad a = \frac{V}{bh}, \quad b = \frac{V}{ah}, \quad h = \frac{V}{ab}$$

Examples.—If the length, breadth and height of a rectangular prism are 2.5, 3.75 and 5 inches, respectively, find the volume, surface and weight.

Thus,

$$V = abh = 2.5 \times 3.75 \times 5 = 46.875 \text{ cubic inches.}$$

$$S = 2h(a + b) = 2 \times 5(2.5 + 3.75)$$
$$= 10 \times 6.25 = 62.5 \text{ square inches.}$$

$$S' = 2h(a + b) + 2ab = 62.5 + 2 \times 2.5 \times 3.75$$
$$= 62.5 + 18.75 = 81.25 \text{ square inches.}$$

If copper weighs .32 pounds per cubic inch, the weight will be,

$$W = .32\,abh = .32 \times 46.875 = 15 \text{ pounds.}$$

If V, b and h are known to be 28, 2.25 and 3, respectively, then

$$a = \frac{V}{bh} = \frac{28}{2.25 \times 3} = \frac{28}{6.75} = 4.15 \text{ inches.}$$

340. Equilateral Triangular Prism.

(D)

$$V = .433a^2h, \quad h = \frac{V}{.433\,a^2}, \quad a = \sqrt{\frac{V}{.433h}}$$

Examples.—Find the volume, surface and weight of a triangular prism, the length of whose sides is 2 inches and height 5 inches.

Thus,

$V = .433a^2h = .433 \times 2 \times 2 \times 5 = 8.66$ cubic inches.

$S = 3ah = 3 \times 2 \times 5 = 30$ square inches.

$S' = 3ah + .866a^2 = 30 + 3.46 = 33.46$ square inches.

If brass weighs .3 pound per cubic inch, the weight will be,

$W = .3 \times .433a^2h = .3 \times 8.66 = 2.59$ pounds.

If $V = 20$ cubic inches and $a = 2.5$ inches,

then

$$h = \frac{V}{.433\,a^2} = \frac{20}{.433 \times 2.5 \times 2.5} = \frac{20}{2.706} = 7.39 \text{ inches.}$$

If $V = 20$ cubic inches and $h = 8$ inches,

then

$$a = \sqrt{\frac{V}{.433h}} = \sqrt{\frac{20}{.433 \times 8}} = \sqrt{\frac{20}{3.464}}$$

$$= \sqrt{5.773} = 2.40 \text{ inches.}$$

341. Hexagonal Prism.

(E)

$$V = 2.598a^2h, \quad h = \frac{V}{2.598a^2}, \quad a = \sqrt{\frac{V}{2.598h}}$$

Examples.—Find the volume, surface and weight of a regular hexagonal prism, whose faces are 1.25 inches and the height 5 inches.

Thus,

$V = 2.598a^2h = 2.598 \times 1.25 \times 1.25 \times 5$
$= 20.29$ cubic inches.

$S = 6ah = 6 \times 1.25 \times 5 = 37.50$ square inches.

If oak wood weighs 49 pounds per cubic foot, the weight will be,

$$W = \frac{49}{1728} \times 20.29 = \frac{994.21}{1728} = .57 \text{ pound.}$$

If $\quad V = 30$ cubic inches and $a = 1.5$ inches, then

$$h = \frac{V}{2.598\,a^2} = \frac{30}{2.598 \times 1.5 \times 1.5} = \frac{30}{5.845} = 5.13 \text{ inches.}$$

342. Circular Cylinder.

(F)

$$V = .7854d^2h, \quad h = \frac{V}{.7854d^2}, \quad d = \sqrt{\frac{V}{.7854h}}$$

Examples.—Find the volume, surface and weight of a circular cylinder 2 inches in diameter and 4 inches high.

Thus,

$$V = .7854d^2h = .7854 \times 2 \times 2 \times 4 = 12.566 \text{ cubic inches.}$$

$$S = 3.1416dh = 3.1416 \times 2 \times 4 = 25.13 \text{ square inches.}$$

$$S' = 3.1416dh + 2 \times .7854d^2 = 25.13 + 2 \times .7854 \times 4$$
$$= 25.13 + 6.28 = 31.41 \text{ square inches.}$$

If maple weighs 44 pounds per cubic foot, its weight will be,

$$W = \frac{44}{1728} \times 12.566 = \frac{552.90}{1728} = .32 \text{ pound.}$$

If $\quad V = 15$ cubic inches and $d = 1.75$ inches,

$$h = \frac{V}{.7854d^2} = \frac{15}{.7854 \times 3.062} = \frac{15}{2.404} = 6.23 \text{ inches.}$$

If $\quad V = 15$ cubic inches and $h = 4$ inches,

$$d = \sqrt{\frac{V}{.7854h}} = \sqrt{\frac{15}{.7854 \times 4}} = \sqrt{\frac{15}{3.1416}}$$
$$= \sqrt{4.774} = 2.18 \text{ inches.}$$

343. Pyramid.

A pyramid is a solid bounded by a number of triangular planes which terminate on one end at a point, called the apex; and on the other end, at the different sides of a polygon, called its base.

The altitude of a pyramid is the perpendicular distance from the apex to the base.

Like the prism, a pyramid may be either right or inclined.

The volume of a pyramid is equal to the area of its base, multiplied by one-third its perpendicular height, or altitude.

The lateral surface of a pyramid is equal to the perimeter of its base, multiplied by one-half its slant height.

A frustum of a prism or pyramid is that part of a prism or pyramid, included between its base and a plane, which is parallel to the base.

A truncated prism or pyramid is that portion of a prism or pyramid, included between its base and a plane, which is oblique to the base.

344. Square Pyramid.

(G)

$$V = \frac{a^2 h}{3}, \quad h = \frac{3V}{a^2}, \quad a = \sqrt{\frac{3V}{h}}$$

Examples.—Find the volume, surface and weight if a equals 3 inches and h equals 5 inches.

Thus, $V = \dfrac{a^2 h}{3} = \dfrac{3 \times 3 \times 5}{3} = \dfrac{45}{3} = 15$ cubic inches.

The slant height $= L = \sqrt{h^2 + \dfrac{a^2}{4}}$

then $S = \dfrac{4aL}{2} = 2a\sqrt{h^2 + \dfrac{a^2}{4}} = 2 \times 3 \sqrt{5 \times 5 + \dfrac{3 \times 3}{4}}$

$= 6\sqrt{25 + 2.25} = 6\sqrt{27.25} = 6 \times 5.22 = 31.32$ square inches.

and $S' = a^2 + 31.32 = 9 + 31.32 = 40.32$ square inches.

If mahogany weighs 53 pounds per cubic foot, find its weight.

$$W = \frac{53}{1728} \times 15 = \frac{795}{1728} = .46 \text{ pound.}$$

Find h when $V = 20$ cubic inches and $a = 2.5$ inches.

Thus, $h = \dfrac{3V}{a^2} = \dfrac{3V}{2.5 \times 2.5} = \dfrac{3 \times 20}{6.25} = 9.60$ inches.

Find a when $V = 20$ cubic inches and $h = 10$ inches.

Thus, $a = \sqrt{\dfrac{3V}{h}} = \sqrt{\dfrac{3 \times 20}{10}} = \sqrt{\dfrac{60}{10}} = \sqrt{6} = 2.45$ inches.

345. Pentagonal Pyramid.

(H)

$$V = .573a^2h, \quad h = \frac{V}{.537a^2}, \quad a = \sqrt{\frac{V}{.573h}}$$

Examples.—Find the volume and weight when $a = 1.5$ and $h = 5$ inches respectively.

$$V = .573a^2h = .573 \times 1.5 \times 1.5 \times 5 = 6.446 \text{ cubic inches.}$$

If the weight of black walnut is 38 pounds per cubic foot, find its weight.

$$W = \frac{38}{1728} \times 6.446 = \frac{38 \times 6.446}{1728} = \frac{244.94}{1728} = .142 \text{ pounds.}$$

346. Circular Cone.

(I)

$$V = .261d^2h, \quad h = \frac{V}{.261d^2}, \quad d = \sqrt{\frac{V}{.261h}}$$

Examples.—Find the volume, surface and weight when $d = 2.5$ inches and $h = 6$ inches.

Thus,

$$V = .261d^2h = .261 \times 2.5 \times 2.5 \times 6 = 9.787 \text{ cubic inches.}$$

$$\text{The slant height} = L = \sqrt{h^2 + \frac{d^2}{4}}$$

then

$$S = 1.570dL = 1.570d\sqrt{h^2 + \frac{d^2}{4}} = 1.570 \times 2.5\sqrt{6^2 + \frac{2.5^2}{4}}$$

$$= 1.570 \times 2.5\sqrt{36 + 1.562} = 3.925\sqrt{37.562}$$

$$= 3.925 \times 6.128 = 24.052 \text{ square inches.}$$

If white pine weighs 40 pounds per cubic foot, its weight in ounces will be,

$$W = \frac{40 \times 16}{1728} \times 9.787 = \frac{640 \times 9.787}{1728} = \frac{6263.680}{1728} = 3.62 \text{ ounces.}$$

Find h when $V = 16$ cubic inches and $d = 3$ inches.

$$h = \frac{V}{.261d^2} = \frac{16}{.261 \times 9} = \frac{16}{2.349} = 6.81 \text{ inches.}$$

Find d when $V = 16$ cubic inches and $h = 7$ inches.

$$d = \sqrt{\frac{V}{.261h}} = \sqrt{\frac{16}{.261 \times 7}} = \sqrt{8.757} = 2.96 \text{ inches.}$$

347. Frustum of Pentagonal Pyramid.

(J)

Let A equal the area of the base and A_1 the area of the parallel face, in square inches, and let h equal the height. The volume may then be found from the equation:

$$V = \frac{h}{3}(A + A_1 + \sqrt{AA_1}) \quad .$$

For a pentagonal pyramid $A = 1.7205a^2$ and $A_1 = 1.7205a_1^2$

Hence

$$V = \frac{h}{3}(1.7205a^2 + 1.7205a_1^2 + \sqrt{1.7205a^2 \times 1.7205a_1^2}$$

Example.—Let $h = 3$, $a = 2.5$ and $a_1 = 2$ inches, respectively, then

$$V = \frac{3}{3}(1.7205 \times 6.25 + 1.7205 \times 4 + \sqrt{1.7205 \times 6.25 \times 1.7205 \times 4}$$

$$= 10.753 + 6.882 + \sqrt{10.753 \times 6.882} = 26.23 \text{ cubic inches.}$$

348. Oblique Prism.

(K)

The volume of any oblique prism is equal to the area of its base multiplied by the altitude. Therefore the equations for the right prisms may be applied to oblique prisms.

349. Regular Truncated Triangular Prism.

(L)

$$\text{Volume} = V = \frac{a^2\sqrt{3}}{4} \times \frac{h + 2h_1}{3}$$

$$= \frac{1.7320}{12} \times a^2(h + 2h_1)$$

$$= .144 \times a^2(h + 2h_1).$$

Example.—Find the volume of a regular triangular truncated prism when $a = 3$, $h = 2$ and $h_1 = 4$ inches, respectively.

Thus, $V = .144 \times a^2(h + 2h_1) = .144 \times 9(2 + 2 \times 4)$
$= .144 \times 9 \times 10 = .144 \times 90 = 12.96$ cubic inches.

350. Sphere.

In terms of the radius,

$$V = \frac{4}{3}\pi r^3 = 4.1888 r^3, \quad r = \sqrt[3]{\frac{V}{4.1888}}$$

In terms of the diameter,

$$V = \frac{1}{6}\pi d^3 = .5236 d^3, \quad d = \sqrt[3]{\frac{V}{.5236}}$$

Examples.—Find the volume of a sphere whose diameter is 6 inches.

Thus,

$$V = 4.1888 r^3 = 4.1888 \times 27 = 113.097 \text{ cubic inches,}$$
or $\quad V = .5236 d^3 = .5236 \times 216 = 113.097 \text{ cubic inches.}$

If cast iron weighs .26 pound per cubic inch, find its weight.

$$W = .26V = .26 \times 113.097 = 29.40 \text{ pounds.}$$

Find the diameter of a sphere whose volume is 100 cubic inches.

Thus,

$$d = \sqrt[3]{\frac{V}{.5236}} = \sqrt[3]{\frac{100}{.5236}} = \sqrt[3]{190.966} = 5.75 \text{ inches.}$$

Find the diameter of a cast-iron sphere whose weight is 50 pounds.

Since $\quad W = 50 = .26V = .26 \times .5236 d^3$

then

$$d = \sqrt[3]{\frac{50}{.26 \times .5236}} = \sqrt[3]{\frac{50}{.1361}} = \sqrt[3]{367.376} = 7.16 \text{ inches.}$$

$$\text{Surface} = S = 4\pi r^2 = 12.5664 r^2, \text{ or } \pi d^2 = 3.1416 d^2$$

Find the surface area of a sphere whose diameter is 7 inches.

$$S = 3.1416 d^2 = 3.1416 \times 7 \times 7 = 153.93 \text{ square inches.}$$

CHAPTER X

REPRODUCTION OF DRAWINGS

351. Introductory.

In all modern construction, such as the building and manufacture of machinery, bridges, buildings, etc., it is necessary to have complete drawings showing the construction in its minutest details. Such drawings are costly to make and after being used in the shop a short time, would become illegible, and thereby destroyed. Therefore it is necessary to make duplicates of the original drawing by one of the various means to be described. The original is retained in the drawing office, which in large manufacturing establishments has a fire-proof vault of some sort for the safe keeping of drawings.

The blue print process is the one most commonly used for reproduction on account of its cheapness and simple manipulation.

352. Making Blue Prints.

Blue prints are made by exposing to the action of light for some time, a specially prepared paper under a tracing or photographic negative; the exposed paper is then washed for some minutes in clear water. After drying, the print without further treatment, is ready for use. These are universally used as working drawings by all kinds of manufacturers and builders. Any number of blue prints can be made from one tracing or negative.

The principal advantage of this method of reproduction of drawings is its extreme cheapness. Some prints, when properly prepared, are of a beautiful deep blue background on which the drawing appears in clear white lines. They are admirably adaptable for shop use because they will not soil as quickly as drawings on white paper.

Prints which are to be used for long periods of time may be mounted on cloth, or cardboard, then varnished with one or two coats of thin shellac. Blue-print paper mounted on cloth may also be obtained from dealers. This is exposed, washed and dried in the same way as the ordinary paper.

It will be observed that since the blue-print process gives white lines on a blue ground from transparent drawings, technically known as dia-positives, it will give blue lines on a white background with a photographic negative. The reason for this is obvious.

Prints which have blue lines on a white ground may be made by first making a blue print on very thin paper, washing and drying in the usual way, then using this as a negative to print from.

If the paper is sufficiently transparent, the white lines will let enough light diffuse; while the blue, if well printed, will be dark enough to prevent the light from penetrating. This method will give blue lines. Theoretically this works well, but practically it gives poor results, as it is difficult to make negatives with sufficient contrast between the blue and white to allow another good print to be made.

If it is necessary to make alterations or corrections on a blue print, the regular drawing instruments or a writing pen may be used. For drawing use a solution of soda and water or quicklime and water. These solutions have a bleaching action and will therefore produce white lines. If an erasure is to be made, use an ordinary blue wax pencil.

353. Exposure of Blue-print Paper.

The time of exposure in bright sunlight will vary from three to five minutes, and on cloudy days will frequently take half an hour. This consumption of time is a serious objection, which in modern practice is overcome by using what is known as a rapid printing paper. This may be obtained at the same price as the ordinary paper. In places where large numbers of blue prints are made, the printing is frequently done with the aid of one or more electric arc lights.

The sensitized paper can be purchased from dealers in drawing materials. It may be obtained by the yard or in 10 or 20 yard rolls, 30, 36, or 42 inches wide.

Dealers will cut any number of yards, or a roll, at a small advance in price, into sheets of any desired size. For drawing-room use it is desirable to have the paper put up into light-proof packages of a dozen or more sheets.

354. Making Blue-print Paper.

The process of making blue-print paper is very simple. A sheet of paper, preferably with a smooth calendered surface, is sensitized, as it is termed, with a solution of citrate of iron and potassium ferricyanide in water.   The proportions are as follows:

Solution A:

Citrate of iron and ammonia................	20 parts
Distilled or rain water.....................	100 parts

Solution B:

Potassium ferricyanide....................	20 parts
Distilled or rain water.....................	100 parts

These chemicals may be purchased at any chemical supply house, and frequently may be had at an ordinary pharmacy. They are very cheap and may be bought in any quantity.

In both solutions, the salts should be thoroughly dissolved before using.   They will keep for a long time in the dark while separated, but will soon decompose after being mixed.

For sensitizing paper, use equal parts of solutions A and B.   It is well to mix only the quantity desired for immediate use.   The paper should be laid upon a clean, smooth table in a dark room illuminated with a red or orange-colored light.   The solution may then be spread on the surface with a wide camels-hair brush, making even strokes both along the length and across the paper, so as to cover the surface as evenly as possible.   It should then be hung up to dry, after which it is ready for use.   Sensitized paper can be preserved for some time, if kept in a dark container, such as a tube or drawer.

The above proportioning of chemicals will make what is known as a slow printing paper.   They may be varied somewhat without any change in results except in time of exposure.

355. More Sensitive Blue-print Paper.

The following formula for quick printing blue-print paper is taken from P. C. Duchochois, *Photographic Reproduction Processes*.   This paper will give good results on an exposure of from 15 to 30 seconds in good sunlight and about 1 minute in the shade.   It is admirably adapted to printing by the aid of electric light.

Solution A:

Tartaric acid.............................	25 parts
Ferric chloride, solution at 45° Baumé.......	90 parts
Water....................................	100 parts

When the acid is dissolved, slowly add 28 per cent. ammonia

to just neutralize the solution. After the solution has become cool, add an equal volume of the following solution:

Solution B:
        Potassium ferricyanide.................... 22 parts
        Water................................. 100 parts

Paper sensitized with this solution will not keep as long as slower printing paper.

356. Printing Frame.

The general construction of printing frames is the same as those used by the photographer. They vary, however, greatly in size. For ordinary drawings a frame 24 inches wide and 30

Fig. 154.—Common blue-printing frame.

inches long will answer very well. It should consist of a strong and stiff frame of wood into which is fitted a sheet of plate glass. It must also have a cover or lid which is divided into three or more sections according to the size of the frame. These are to be hinged together and each section should be provided with a rather stiff steel or brass spring. The frame should be provided with projections suitable for holding the ends of the springs, which in turn press the cover firmly to the glass plate. It is desirable to have a thick piece of felt the same size as the cover, which should be placed on the paper under the cover. This will insure that perfect contact between the tracing and the paper, which is necessary for producing clear and unbroken lines on the blue print.

The method of using the printing frame is as follows: Place the tracing with the ink-lined surface on the glass, then the blue-print paper with sensitized surface on the tracing, then put on the felt and cover and clamp it down with the springs. The sensitized paper is then ready for exposure.

Fig. 154 shows an 18 × 24 inch printing frame. The cover is cut into three parts, one part being opened to show the glass, tracing, sensitized paper and felt.

357. Printing Apparatus.

In places where large numbers of blue prints are made, the printing apparatus for rapid work consists of a plate-glass cylinder about 30 inches in diameter and 60 inches high. The cylinder is open at both ends, the lower end being mounted on an iron frame. This frame is made to revolve by a suitable mechanism and at the same time an electric arc light moves up and down inside the cylinder. The tracing, sensitized paper, and proper backing are wound around the cylinder and fastened. An exposure is complete after the light has moved down to the bottom of the cylinder and up again. The speed of revolution may be varied by the operator at will. The apparatus is very costly.

358. Hectography.

When a small number of reproductions of moderate size are wanted, hectography may be used. This consists of a semi-hard pad which will take impressions from a specially prepared drawing. The pad may be made by the following formula:

Glycerine	12 ounces
Glue	3 ounces
Sugar	3 ounces
Kaolin	5 ounces
Water	12 ounces

Soak the glue in the water and when soft gently heat it until melted; then add the glycerine, sugar and kaolin, stir the mass, and while still warm pour it into a shallow pan about 8 inches wide and 12 inches long.

Place this on a level surface, in a room free from dust, to cool. When cold and quite hard, the pad is ready for use.

359. The drawing to be copied must be made on a smooth paper which has a hard calendered surface, and with ink commercially known as hectograph ink, which may be obtained in blue, purple, red, or black colors. It may also be made by the following formula:

Aniline color	1	part
Alcohol	1	part
Glycerine	.25	part
Water	6	parts

Dissolve the color in the alcohol, add the glycerine and water and stir the mixture gently, adding a few drops carbolic acid as a preservative. It is then ready for use.

The ink may be obtained from dealers in drawing materials at a very low cost and is, therefore, not generally made by the draftsman.

The hectograph has the advantage of yielding reproductions in colors. It is not used very extensively for large drawings, but for small, simple sketches of a limited number, it will answer very well.

360. For use moisten the pad slightly with water, and when it is nearly dry, lay the drawing to be copied with the inked surface on the pad and gently press it down to insure perfect contact. Let it remain for about 1 minute, then take hold of one corner of the drawing and slowly remove it. A reverse impression is now on the pad. To obtain copies, lay on, and gently press down, a sheet of blank paper and remove by taking hold of one corner. About thirty or forty good copies may be made in this way. If the pad is to be used again immediately, the ink still remaining may be removed by washing the surface with a sponge. If it is not removed in this way, it will, after a time, be entirely absorbed and will not interfere with other impressions.

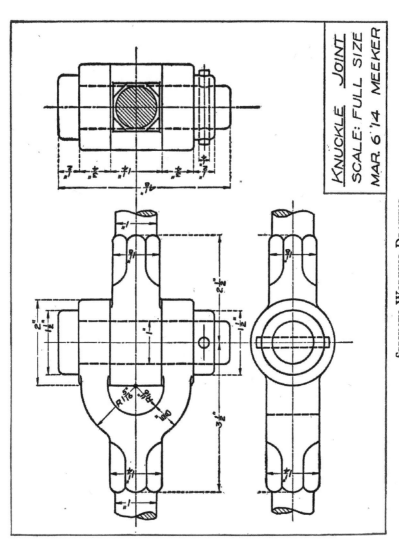

KNUCKLE JOINT
SCALE: FULL SIZE
MAR. 6 '14   MEEKER

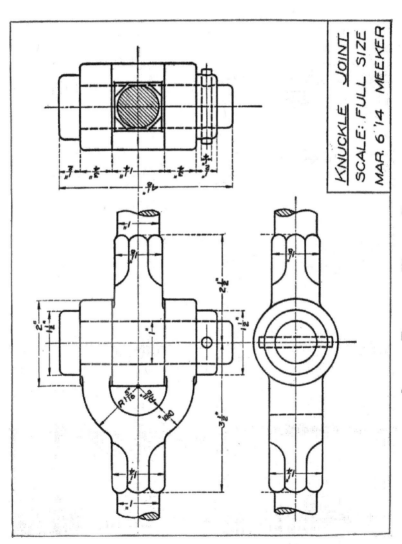

KNUCKLE JOINT
SCALE: FULL SIZE
MAR. 6 '14   MEEKER

SPECIMEN TRACING OF WORKING DRAWING

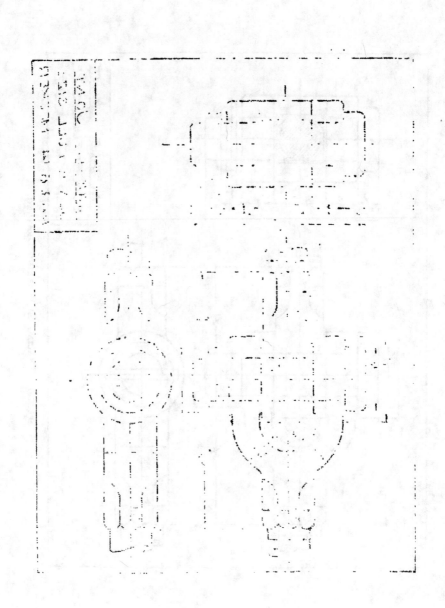

INDEX

www.ingramcontent.com/pod-product-compliance
Lightning Source LLC
LaVergne TN
LVHW012204040326
832903LV00003B/108